PHYTOHORMONES - SIGNALING MECHANISMS AND CROSSTALK IN PLANT DEVELOPMENT AND STRESS RESPONSES

Edited by **Mohamed El-Esawi**

Phytohormones - Signaling Mechanisms and Crosstalk in Plant Development and Stress Responses
http://dx.doi.org/10.5772/65234
Edited by Mohamed El-Esawi

Contributors

Mohamed Ahmed El-Esawi, Masayuki Fujita, Mirza Hasanuzzaman, Kamrun Nahar, Tasnim Farha Bhuiyan, Taufika Islam Anee, Masashi Inafuku, Hirosuke Oku, Agata Daszkowska-Golec, Anna Skubacz, Marek Marzec, Fahim Nawaz, Rana Nauman Shabbir, Muhammad Shahbaz, Sadia Majeed, Muhammad Raheel, Waseem Hassan, Muhammad Amir Sohail, Francisca Parada, Patricio Arce-Johnson, Carmen Espinoza, Ales Marek

Published by InTech

Janeza Trdine 9, 51000 Rijeka, Croatia

Publishing Process Manager Nina Kalinic
Technical Editor SPi Global
Cover InTech Design team

First published July, 2017
Printed in Croatia
Legal deposit, Croatia: National and University Library in Zagreb

Additional hard copies can be obtained from orders@intechopen.com

Phytohormones - Signaling Mechanisms and Crosstalk in Plant Development and Stress Responses , Edited by Mohamed El-Esawi
p. cm.
Print ISBN 978-953-51-3411-4
Online ISBN 978-953-51-3412-1

For EU product safety concerns:
IN TECH d.o.o., Prolaz Marije Krucifikse Kozulić 3, 51000 Rijeka, Croatia,
info@intechopen.com or visit our website at intechopen.com.

Meet the editor

Dr. Mohamed El-Esawi is currently a visiting research fellow at the University of Cambridge in the United Kingdom and an assistant professor of Molecular Biology and Genetics at Botany Department of Tanta University in Egypt. Dr. El-Esawi received his BSc and MSc degrees from Tanta University and his PhD degree in Plant Genetics and Molecular Biology from Dublin Institute of Technology in Ireland. Afterwards, Dr. El-Esawi joined the University of Warwick in the UK, University of Sorbonne (Paris VI) in France and University of Leuven (KU Leuven) in Belgium as a visiting research fellow. His research focuses on plant genetics, genomics, molecular biology, molecular physiology, developmental biology, plant-microbe interaction and bioinformatics. He has authored several international journal articles and book chapters and participated at more than 60 conferences and workshops worldwide. Dr. El-Esawi is currently involved in several research projects on biological sciences.

Contents

Preface

Phytohormones are regulatory compounds that play crucial roles in plant development and comprise five main groups, namely, auxins, cytokinins, abscisic acid, gibberellins and ethylene. Salicylates, jasmonates, brassinosteroids, strigolactones, polyamines and some peptides were known as new families of phytohormones. Auxins, gibberellins, cytokinins, brassinosteroids, strigolactones and polyamines play a major role in plant growth and development, whereas ethylene, abscisic acid, salicylates and jasmonates have important roles in stress responses. Hormone actions form a signaling network and mutually regulate various systems in plants.

This book brings together recent work and progress that has recently been made in the dynamic field of phytohormone regulation in plant development and stress responses. It also provides new insights and sheds new light regarding the exciting hormonal cross talk phenomenon in plants. This book will provoke interest in many readers and scientists, who can find this information useful for the advancement of their research works.

The book includes seven chapters. The first chapter "Hormonal Regulation in Plant Development and Stress Tolerance" presents an overview and discusses recent progresses of phytohormones' roles and their cross talk in plant development and stress tolerance. The second chapter "Recent Developments in a Radio-labeling of Brassinosteroids" provides a comprehensive overview on methodologies used for radio-labeling of brassinosteroids as one of the newest class of phytohormones. The third chapter "Salicylic Acid: An All-Rounder in Regulating Abiotic Stress Responses in Plants" sheds light on the recent progress on the regulatory role of SA in mitigating abiotic stress. The fourth chapter "Seed Dormancy: The Complex Process Regulated by Abscisic Acid, Gibberellins and Other Phytohormones That Makes the Seed Germination Work" describes recent advances in understanding the complex process of seed dormancy, regulated by many phytohormonal pathways and their components. The fifth chapter "Strigolactone Signaling in Plants" summarizes the recent discoveries in the signal transduction pathway of SLs and describes the model of SL perception and signaling. The sixth chapter "Cross Talk between Nitric Oxide and Phytohormones Regulates Plant Development During Abiotic Stresses" discusses the current state of knowledge of cross talk between signaling pathways of NO and phytohormones in plants exposed to various abiotic stresses. The seventh chapter "Phytohormonal Control over the Grapevine Berry Development" discusses phytohormones, *their sig*naling pathways and their association to berry development in Vitis vinifera.

The book editor would like to thank Ms. Nina Kalinić, Publishing Process Manager, for her wholehearted cooperation in the publication of this book.

Mohamed Ahmed El-Esawi, PhD
Sainsbury Laboratory, University of Cambridge,
Cambridge, United Kingdom

Botany Department, Faculty of Science,
Tanta University, Egypt

Chapter 1

Introductory Chapter: Hormonal Regulation in Plant Development and Stress Tolerance

Mohamed A. El-Esawi

Additional information is available at the end of the chapter

http://dx.doi.org/10.5772/intechopen.69806

1. Introduction

Phytohormones are regulatory compounds produced in low concentrations and serve as chemical messengers to regulate various plant physiological and developmental processes. They also play essential roles in signal transduction pathways during stress response and regulate internal and external stimuli [1]. Phytohormones comprise five main groups, namely auxins (IAAs), cytokinins (CKs), abscisic acid (ABA), gibberellins (GAs) and ethylene (ET). Salicylates (SAs), jasmonates (JAs), brassinosteroids (BRs), strigolactones (SLs), polyamines, and some peptides represent new families of phytohormones.

Hormone actions form a signaling network and regulate various systems in plants. The interacting actions among hormone signal transduction cascades are called crosstalk [2, 3]. Phytohormones interact by activating a phosphorylation cascade or a common second messenger. Furthermore, several phytohormones interact together, forming a defense network against environmental stresses such as JA, SA, and ABA which play a crucial role in regulating signaling pathways [3]. Understanding the crosstalk between these phytohormones and defense signaling pathways helps reveal new important targets for developing host resistance mechanisms [3, 4]. Here, the current work presents an overview and discusses recent progresses of phytohormone roles and their crosstalk in plant development and stress tolerance.

2. Phytohormones signaling roles

2.1. Abscisic acid

Abscisic acid (ABA), termed stress hormone, plays an important role in plant leaves abscission and abiotic stresses tolerance [3]. ABA also has important roles in various plant developmental and physiological processes such as seed dormancy, embryo morphogenesis, stomatal opening, cell turgor maintenance, and biosynthesis of lipids and storage proteins [3, 5]. ABA regulates protein-encoding genes [6]. ABA enables plants to survive under severe environmental factors [7] and water-deficit conditions [8]. ABA is also important for root growth and architectural modifications under nitrogen deficiency [9] and drought stress [10]. Furthermore, ABA is involved in the biosynthesis of dehydrins, osmoprotectants and protective proteins [3, 11, 12].

2.2. Auxins

Some pathways for auxin (IAA) biosynthesis in plants have been reported so far including one tryptophan-independent and four tryptophan-dependent pathways [3, 13]. IAA plays an important role in plant growth and development as well as in regulating growth under stress factors [14]. IAA plays essential roles in plant adaptation to salinity [15] and heavy metal stresses [16]. Furthermore, auxins induce the transcription of the primary auxin response genes which are identified in various plants such as rice, *Arabidopsis* and soybean [3, 17]. Auxin also regulates crosstalk between biotic sand abiotic stresses [18].

2.3. Cytokinins

Cytokinins (CKs) regulate plant growth and development [3, 19]. They are also involved in abiotic stresses [20] such as salinity [21] and drought [19]. They are also important for various crop traits such as productivity and enhanced stress tolerance [3, 22]. CKs also release seeds from dormancy [18] and are considered as abscisic acid antagonists [23]. Decreased CK content promotes apical dominance, which assists in the adaptation to drought stress [3, 20].

2.4. Ethylene

Ethylene (ET) is a gaseous phytohormone regulating plant growth and developmental processes, including flower senescence, fruit ripening, and petal and leaf abscission, as well as regulating stress responses [3, 24, 25]. Ethylene biosynthesis begins from methionine via S-adenosyl-L-methionine and the cyclic amino acid ACC. ACC synthase converts S-adenosyl-L-methionine to ACC, whereas ACC oxidase catalyzes the conversion of ACC to ET. Various abiotic stresses affect endogenous ethylene levels in plant species. Higher ET concentrations promote stress tolerance [26]. Ethylene may combine with other hormones such as jasmonates and salicylic acid and plays crucial roles in regulating plant defense against biotic stress factors [3, 1]. Ethylene and abscisic acid may act together to regulate plant growth and development [3].

2.5. Gibberellins

Gibberellins (GAs) are carboxylic acids that may regulate plant growth and development [27]. They positively regulate leaf expansion, seed germination, stem elongation, flower development and trichome initiation [3, 28]. They also play important role in abiotic stress tolerance [29] such as osmotic stress. GAs may interact with other hormones and regulate various developmental processes [30]. These interactions may involve both negative and positive regulatory roles [3, 30].

2.6. Brassinosteroids

Brassinosteroids (BRs) comprise polyhydroxy steroidal phytohormones which regulate plant growth and developmental processes including root and stem growth, and flower initiation and development [3]. BRs were first isolated from *Brassica napus*. Brassinolide, 24-epibrassinolide, and 28-homobrassinolide are the most bioactive BRs widely used in physiological studies [31]. They are found in flower buds, pollen, fruits, vascular cambium, seeds, leaves, roots, and shoots [32]. BRs also play important roles in abiotic stress responses such as chilling, high temperature, soil salinity, drought, light, flooding, and organic pollutants [3].

2.7. Jasmonates

Jasmonates (JAs) are multifunctional phytohormones derived from the membrane fatty acids metabolism and are widely distributed in several plant species [3]. JAs play crucial roles in growth and developmental processes such as fruiting, flowering, senescence and secondary metabolism [3, 33]. JAs are also involved in biotic and abiotic stress responses such as salinity, drought, irradiation and low temperature [3, 34]. Exogenous concentrations of methyl jasmonate (MeJA) minimize salinity stress symptoms [35]. Additionally, endogenous levels of JA are induced in roots under salinity stress [36]. JA levels also reduce heavy metal stress through inducing the antioxidant machinery [3, 37]. MeJA accumulates phytochelatins, conferring tolerance against Cu and Cd stress [38].

2.8. Salicylic acid

Salicylic acid (SA) is a phenolic compound which regulates the expression of pathogenesis-associated proteins [39]. SA plays an important role in plant growth and development, as well as in biotic and abiotic stress responses [3, 40]. SA has two biosynthesis pathways: the major isochorismate (IC) pathway and the phenylalanine ammonia-lyase (PAL) pathway. Low levels of SA promote the plant antioxidant capacity [3]. However, the high SA levels may result in cell death [41]. SA comprises genes encoding chaperones, antioxidants, heat shock proteins, and secondary metabolite biosynthetic genes such as cinnamyl alcohol dehydrogenase, sinapyl alcohol dehydrogenase and cytochrome P450 [3, 41]. SA may also combine with ABA to regulate drought response [39]. However, the SA mechanism in abiotic stress tolerance remains mainly unknown and still needs more investigations.

2.9. Strigolactones

Strigolactones (SLs) are carotenoid-derived compounds, produced in small quantities in roots, or synthesized in several plant species [3, 42]. SLs play an important role in root architecture and development [43]. They induce nodulation during interaction processes [44] and may be used for inducing the parasitic plants seed germination [45]. SLs are also involved in biotic and abiotic responses [3].

3. Phytohormones crosstalk

Sessile plants should maintain growth plasticity and adaptation ability to severe environmental conditions. Stress-responsive hormones assist in the alteration of cellular dynamics and thus regulating plant growth under stress conditions [3, 46]. The interacting actions among hormone signal transduction cascades are called crosstalk and form a signaling network [2, 3]. In this case, hormones interact by activating a phosphorylation cascade or a common second messenger. Several phytohormones interact together forming a defense network against environmental stresses such as JA, SA, and ABA which play a crucial role in regulating signaling pathways [3]. Understanding the crosstalk between these phytohormones and defense signaling pathways helps reveal new important targets for developing host resistance mechanisms [3, 4].

A complex signaling network regulates stomatal closure. ABA regulates gene expression which mediates root growth maintenance and water uptake. ABA interacts with signaling molecules and other phytohormones such as nitric oxide and JA to induce stomatal closure, as well as to induce genes controlling response to cytokinin, ethylene or auxin [2, 3]. Furthermore, exogenous treatment of ABA down-regulated the key cytokinin biosynthetic pathway gene, termed isopentenyltransferase, but up-regulated genes encoding cytokinin dehydrogenases and oxidases [3, 21]. GA is also included in the hormonal crosstalk in environmental signals [47]. In conclusion, like the potential use of molecular and genetic markers in crop improvement [48–57], phytohormones play crucial roles in development and stress tolerance of crops.

Author details

Mohamed A. El-Esawi

Address all correspondence to: mohamed.elesawi@science.tanta.edu.eg

1 Botany Department, Faculty of Science, Tanta University, Tanta, Egypt

2 The Sainsbury Laboratory, University of Cambridge, Cambridge, United Kingdom

References

[1] Kazan K. Diverse roles of jasmonates and ethylene in abiotic stress tolerance. Trends in Plant Science. 2015;**20**:219-229

[2] Harrison MA. Cross-talk between phytohormone signaling pathways under both optimal and stressful environmental conditions. In: Khan NA, Nazar R, Iqbal N, Anjum NA, editors. Phytohormones and Abiotic Stress Tolerance in Plants. Berlin, Heidelberg: Springer-Verlag; 2012. pp. 49-76

[3] Wani SH, Kumar V, Shriram V, Sah SK. Phytohormones and their metabolic engineering for abiotic stress tolerance in crop plants. The Crop Journal. 2016;**4**:162-176

[4] Grant MR, Jones JD. Hormone (dis)harmony moulds plant health and disease. Science. 2009;**324**:750-752

[5] Sreenivasulu N, Radchuk V, Alawady A, Borisjuk L, Weier D, Staroske N, Fuchs J, Miersch O, Strickert M, Usadel B, Wobus U, Grimm B, Weber H, Weschke W. De-regulation of abscisic acid contents causes abnormal endosperm development in the barley mutant *seg8*. The Plant Journal. 2010;**64**:589-603

[6] Nemhauser JL, Hong F, Chory J. Different plant hormones regulate similar processes through largely non overlapping transcriptional responses. Cell. 2006;**126**:467-475

[7] Keskin BC, Sarikaya AT, Yuksel B, Memon AR. Abscisic acid regulated gene expression in bread wheat. Australian Journal of Crop Science. 2010;**4**:617-625

[8] Wilkinson S, Kudoyarova GR, Veselov DS, Arkhipova TN, Davies WJ. Plant hormone interactions: Innovative targets for crop breeding and management. Journal of Experimental Botany. 2012;**63**:3499-3509

[9] Zhang S, Hu J, Zhang Y, Xie XJ, Knapp A. Seed priming with brassinolide improves lucerne (*Medicago sativa* L.) seed germination and seedling growth in relation to physiological changes under salinity stress. Australian Journal of Agricultural Research. 2007;**58**:811-815

[10] Giuliani S, Sanguineti MC, Tuberosa R, Bellotti M, Salvi S, Landi P. Root-ABA1 a major constitutive QTL affects maize root architecture and leaf ABA concentration at different water regimes. Journal of Experimental Botany. 2005;**56**:3061-3070

[11] Verslues PE, Agarwal M, Katiyar-Agarwal S, Zhu J, Zhu JK. Methods and concepts in quantifying resistance to drought, salt and freezing, abiotic stresses that affect plant water status. The Plant Journal. 2006;**45**:523-539

[12] Sreenivasulu N, Harshavardhan VT, Govind G, Seiler C, Kohli A. Contrapuntal role of ABA: Does it mediate stress tolerance or plant growth retardation under long-term drought stress? Gene. 2012;**506**:265-273

[13] Mano Y, Nemoto K. The pathway of auxin biosynthesis in plants. Journal of Experimental Botany. 2012;**63**:2853-2872

[14] Kazan K. Auxin and the integration of environmental signals into plant root development. Annals of Botany. 2013;**112**:1655-1665

[15] Fahad S, Hussain S, Bano A, Saud S, Hassan S, Shan D, Khan FA, Khan F, Chen YT, Wu C, Tabassum MA, Chun MX, Afzal M, Jan A, Jan MT, Huang JL. Potential role of phytohormones and plant growth-promoting rhizobacteria in abiotic stresses: Consequences for changing environment. Environmental Science and Pollution Research. 2015;**22**:4907-4921

[16] Egamberdieva D. Alleviation of salt stress by plant growth regulators and IAA producing bacteria in wheat. Acta Physiologiae Plantarum. 2009;**31**:861-864

[17] Javid MG, Sorooshzadeh A, Moradi F, Sanavy SA, Allahdadi I. The role of phytohormones in alleviating salt stress in crop plants. Australian Journal of Crop Science. 2011;**5**:726-734

[18] Fahad S, Hussain S, Matloob A, Khan FA, Khaliq A, Saud S, Hassan S, Shan D, Khan F, Ullah N, Faiq M, Khan MR, Tareen AK, Khan A, Ullah A, Ullah N, Huang JL. Phytohormones and plant responses to salinity stress: A review. Plant Growth Regulation. 2015;**75**:391-404

[19] Kang NY, Cho C, Kim NY, Kim J. Cytokininreceptor-dependent and receptor-independent pathways in the dehydration response of *Arabidopsis thaliana*. Journal of Plant Physiology. 2012;**169**:1382-1391

[20] O–Brien JA, Benkova E. Cytokinin cross-talking during biotic and abiotic stress responses. Frontiers in Plant Science. 2013;**4**:451. DOI: 10.3389/fpls.2013.00451

[21] Nishiyama R, Watanabe Y, Fujita Y, Tien LD, Kojima M, Werner T, Vankova R, Yamaguchi-Shinozaki K, Shinozaki K, Kakimoto T, Sakakibara H, Schmuelling T, Lam-Son PT. Analysis of cytokinin mutants and regulation of cytokinin metabolic genes reveals important regulatory roles of cytokinins in drought, salt and abscisic acid responses, and abscisic acid biosynthesis. Plant Cell. 2011;**23**:2169-2183

[22] Zalabák D, Pospíšilová H, Šmehilová M, Mrízová K, Frébort I, Galuszka P. Genetic engineering of cytokinin metabolism: Prospective way to improve agricultural traits of crop plants. Biotechnology Advances. 2013;**31**:97-117

[23] Pospíšilová J. Participation of phytohormones in the stomatal regulation of gas exchange during water stress. Biologia Plantarum. 2003;**46**:491-506

[24] Gamalero E, Glick BR. Ethylene and abiotic stress tolerance in plants. In: Ahmed P, Prasad MNV, editors. Environmental Adaptations and Stress Tolerance of Plants in the Era of Climate Change. New York: Springer; 2012. pp. 395-412

[25] Groen SC, Whiteman NK. The evolution of ethylene signaling in plant chemical ecology. Journal of Chemical Ecology. 2014;**40**:700-716

[26] Shi Y, Tian S, Hou L, Huang X, Zhang X, Guo H, Yang S. Ethylene signaling negatively regulates freezing tolerance by repressing expression of *CBF* and Type-A *ARR* genes in *Arabidopsis*. Plant Cell. 2012;**24**:2578-2595

[27] Sponsel VM, Hedden P. Gibberellin, biosynthesis and inactivation. In: Davies PJ, editor. Plant Hormones Biosynthesis, Signal Transduction, Action! Dordrecht: Springer; 2004. pp. 63-94

[28] Yamaguchi S. Gibberellin metabolism and its regulation. Annual Review of Plant Physiology. 2008;**59**:225-251

[29] Colebrook EH, Thomas SG, Phillips AL, Hedden P. The role of gibberellin signaling in plant responses to abiotic stress. Journal of Experimental Biology. 2014;**217**:67-75

[30] Munteanu V, Gordeev V, Martea R, Duca M. Effect of gibberellin cross talk with other phytohormones on cellular growth and mitosis to endoreduplication transition. International Journal of Advanced Research in Biological Sciences. 2014;**1**(6):136-153

[31] Vardhini BV, Anuradha S, Rao SS. Brassinosteroids: A great potential to improve crop productivity. Indian Journal of Plant Physiology. 2006;**11**:1-12

[32] Bajguz A, Hayat S. Effects of brassinosteroids on the plant responses to environmental stresses. Plant Physiology and Biochemistry. 2009;**47**:1-8

[33] Fahad S, Nie L, Chen Y, Wu C, Xiong D, Saud S, Hongyan L, Cui K, Huang J. Crop plant hormones and environmental stress. Sustainable Agriculture Reviews. 2015;**15**:371-400

[34] Pauwels L, Inze D, Goossens A. Jasmonate-inducible gene: What does it mean? Trends in Plant Science. 2009;**14**:87-91. DOI: 10.1016/j.tplants.2008.11.005

[35] Yoon J, Hamayun M, Lee SK, Lee IJ. Methyl jasmonate alleviated salinity stress in soybean. Journal of Crop Science and Biotechnology. 2009;**12**:63-68

[36] Wang Y, Mopper S, Hasentein KH. Effects of salinity on endogenous ABA, IAA, JA, and SA in *Iris hexagona*. Journal of Chemical Ecology. 2001;**27**:327-342

[37] Yan Z, Chen J, Li X. Methyl jasmonate as modulator of Cd toxicity in *Capsicum frutescens* var. *fasciculatum* seedlings. Ecotoxicology and Environmental Safety. 2013;**98**:203-209

[38] Maksymiec W, Wojcik M, Krupa Z. Variation in oxidative stress and photochemical activity in *Arabidopsis thaliana* leaves subjected to cadmium and excess copper in the presence or absence of jasmonate and ascorbate. Chemosphere. 2007;**66**:421-427

[39] Miura K, Tada Y. Regulation of water, salinity, and cold stress responses by salicylic acid. Frontiers in Plant Science. 2014;**5**:4. DOI: 10.3389/fpls.2014.00004

[40] Rivas-San VM, Plasencia J. Salicylic acid beyond defense: Its role in plant growth and development. Journal of Experimental Botany. 2011;**62**:3321-3338

[41] Jumali SS, Said IM, Ismail I, Zainal Z. Genes induced by high concentration of salicylic acid in *Mitragyna speciosa*. Australian Journal of Crop Science. 2011;**5**:296-303

[42] Koltai H, Beveridge CA. Strigolactones and the coordinated development of shoot and root. In: Baluska F, editor. Long-Distance Systemic Signaling and Communication in Plants. Berlin: Springer; 2013. pp. 189-204

[43] Ruyter-Spira C, Kohlen W, Charnikhova T, van Zeijl A, van Bezouwen L, de Ruijter N, Cardoso C, Lopez-Raez JA, Matusova R, Bours R, Verstappen F, Bouwmeester H. Physiological effects of the synthetic strigolactone analog GR24 on root system architecture in *Arabidopsis*: Another belowground role for strigolactones? Plant Physiology. 2011;**155**:721-734

[44] Foo E, Davies NW. Strigolactones promote nodulation in pea. Planta. 2011;**234**:1073-1081

[45] Vurro M, Yoneyama K. Strigolactones-intriguing biologically active compounds: Perspectives for deciphering their biological role and for proposing practical application. Pest Management Science. 2012;**68**:664-668

[46] Kohli A, Sreenivasulu N, Lakshmanan P, Kumar PP. The phytohormone crosstalk paradigm takes center stage in understanding how plants respond to abiotic stresses. Plant Cell Reports. 2013;**32**:945-957

[47] Depuydt S, Hardtke CS. Hormone signalling crosstalk in plant growth regulation. Current Biology. 2011;**21**:R365-R373

[48] Consentino L, Lambert S, Martino C, Jourdan N, Bouchet PE, Witczak J, Castello P, El-Esawi M, Corbineau F, d'Harlingue A, Ahmad M. Blue-light dependent reactive oxygen species formation by *Arabidopsis* cryptochrome may define a novel evolutionarily conserved signaling mechanism. New Phytologist. 2015;**206**:1450-1462

[49] El-Esawi MA. Genetic diversity and evolution of *Brassica* genetic resources: From morphology to novel genomic technologies—A review. Plant Genetic Resources. C. 2016:1-12. DOI:10.1017/S1479262116000058

[50] El-Esawi MA. Micropropagation technology and its applications for crop improvement. In: Anis M, Ahmad N,editors. Plant Tissue Culture: Propagation, Conservation and Crop Improvement. Springer Singapore; 2016b. pp. 523-545

[51] El-Esawi MA. Nonzygotic embryogenesis for plant development. In: Anis M, Ahmad N, editors. Plant Tissue Culture: Propagation, Conservation and Crop Improvement. Springer Singapore; 2016. pp. 583-598

[52] El-Esawi MA. Somatic hybridization and microspore culture in *Brassica* improvement. In: Anis M, Ahmad N, editors. Plant Tissue Culture: Propagation, Conservation and Crop Improvement. Springer Singapore; 2016. pp. 599-609

[53] El-Esawi MA, Sammour R. Karyological and phylogenetic studies in the genus *Lactuca* L. (Asteraceae). Cytologia. 2014;**79**:269-275

[54] El-Esawi M, Glascoe A, Engle D, Ritz T, Link J, Ahmad M. Cellular metabolites modulate *in vivo* signaling of *Arabidopsis* cryptochrome-1. Plant Signaling & Behavior. 2015;**10**:e1063758

[55] El-Esawi MA, Germaine K, Bourke P, Malone R. Genetic diversity and population structure of *Brassica oleracea* germplasm in Ireland using SSR markers. Comptes Rendus Biologies. 2016;**339**:133-140

[56] El-Esawi MA, Germaine K, Bourke P, Malone R. AFLP analysis of genetic diversity and phylogenetic relationships of *Brassica oleracea* in Ireland. Comptes Rendus Biologies. 2016;**339**:163-170

[57] Jourdan N, Martino C, El-Esawi M, Witczak J, Bouchet PE, d'Harlingue A, Ahmad M. Blue-light dependent ROS formation by *Arabidopsis*cryptochrome-2 may contribute towards its signaling role. Plant Signaling & Behavior. 2015;**10**:e1042647

Chapter 2

Recent Developments in a Radio-labeling of Brassinosteroids

Aleš Marek

Additional information is available at the end of the chapter

http://dx.doi.org/10.5772/intechopen.68584

Abstract

The chapter provides a comprehensive overview on methodologies used for radio-labeling of brassinosteroids as one of the newest class of phytohormones. Discussed labeling strategies are lined up in terms of reached specific activities (SA) of brassinosteroids (BRs) as a key parameter for further utilization of such labeled drugs. The chapter is focused on two key natural radio-isotopes (tritium and carbon-14) used for drug tracing in pharmaceutical research.

Keywords: brassinosteroids, radio-isotope labeling, tritium, carbon-14

1. Introduction

1.1. Radio-labeling

Radioactive labels used for tracing of studied ligands have long been a part of the biological laboratory repertoire. Radioactivity gives a clear, unmistakable signal, and its use is fairly straightforward. Because of smooth traceability, visualization in organ tissues, quantification (liquid scintillation counting [LSC]), and unsurpassed sensitivity of radio-labeled molecules the radio-labeling is a powerful and practical tool to closely follow accurate mass balance and monitor the fate of a molecule on the molecular level and its biochemically transformed derivatives. Pharmacokinetic studies have traditionally used radio-labeled target compounds as a means for evaluating body absorption, distribution, metabolism, and excretion (ADME) [1, 2]. The use of radioligands is essential tool in binding assays aimed at ligand–receptor structure–activity relationship studies, which however requires high-specific activity (SA—a qualitative parameter) of studied ligands (because of, in general, very low concentration of receptors in tissues). Weak β-emitters such as tritium (^3H, T) and carbon-14 (^{14}C) are by far the most versatile and convenient natural labels available [3]. These two isotopes preserve molecular structure (no added tags or pendant groups that alter or change the structure). The advantages of ^3H compared to ^{14}C are much higher specific activity, significantly lower cost of starting material, and environment friendly radioactive waste management (shorter half time, **Table 1**). Also, an introduction of radio-isotope in a later stage of synthetic sequence (often in the last step) is a critical benefit in terms of synthetic yield, safety handling, and waste disposal. In general, all above-mentioned issues come out in favor of tritium over carbon-14. On the other hand, a relevant advantage of carbon-14 is lower potential of label loss. Label selection is usually at the discretion of the investigator and studies can be reported using either ^3H or ^{14}C label. For instance, tritium label could be applied to earlier stage development studies and then switched to a ^{14}C label for the later stage development studies, for example, an advanced human ADME [4]. Each compound radio-labeled at the non-exchangeable and metabolically stable position need to possess radiochemical purity (RCP) basically over 97%. Instability of all radioligands caused by self-radiolysis requires a need to check a radiochemical purity of studied radio-labeled material before a particular experiment is carried out. Such instability can be significantly suppressed by appropriate storage of labeled material. In general, samples stored at −196°C (Cryoflex-sealed vial immersed in liquid nitrogen) in alcohol-reached medium last over 1–2 years in acceptable quality. Radiochemical purity (RCP) is then usually still over 95%. Samples in such conditions can be often used immediately for biological experiments and no further purification is needed. On the contrary, radio-labeled drugs

	Tritium	Carbon-14
Radioactive half-life	12.33 years	5730 years
Specific activity–labeled drugs (1 atom per molecule)	1.066 TBq/mmol	2.309 GBq/mmol
	29.1 Ci/mmol	0.0624 Ci/mmol
Specific activity—element	3.56×10^{14} Bq/g	1.66 × 1011 Bq/g
	2.57 Ci/mL	
Type of radiation (emission probability, %)	β^- (100%)	β^- (100%)
Energy	E_{max} = 18.6 keV E_{avg} = 5.7 keV	E_{max} = 156 keV E_{avg} = 49 keV
Maximum penetration air/water(tissue)/glass	6 mm/6 μm/2 μm	24 cm/0.250 um/170 μm
Decay product	$^3He^+$ (stable)	$^{14}N^+$ (stable)
Detection and measurement	LSC (undetectable by portable survey meters)	LSC Geiger-Mueller [10% efficiency]
Shielding	None required—not an external radiation hazard	None required—mCi quantities not an external radiation hazard

Table 1. Nuclear characteristics of discussed radio-isotopes.

stored in a refrigerator (+4°C) over 1 year rarely show better than 90% RCP. Foremost advantage of using radio-labeled drugs is that the radioactivity is easily detected and quantified using liquid scintillation techniques in a very low limit of detection (technically <1 Bq/L) [5, 6].

1.2. Brassinosteroids: a newest class of phytohormones

The entire evolutionary process in plants is regulated by the changes of hormonal concentration, tissue sensitivity, and their interaction during the entire life cycle of plants. One of the most recent groups of phytohormones represents brassinosteroids. They occur at low levels distributed throughout the plant kingdom [7]. The ability of plants to biosynthesize a large variety of such steroids were discovered in 1970 by Mitchell et al. who first from 40 kg of bee-collected rape pollen of *Brassica napus L.* isolated and characterized a steroidal lactone named brassinolide [8]. Few years later, castasterone was found in insect galls of *Castanea crenata spp.* Up to date, more than 70 structurally and functionally related polyhydroxylated sterol derivatives are known and this group of compounds has been identified as the natural brassinosteroids (BRs) [9]. All BRs have a 5α-cholestane skeleton, with functional variations due to differences in orientations of oxygenated functions on the skeleton [10, 11].

The highest concentration of BRs in plants is detected in the reproductive organs (pollen and seeds; 1–100 ng/g). They were also detected in other plant organs from roots to leaves. They are involved in various kinds of regulatory actions on growth and development, that is, stimulation of cell expansion, cell division, stress tolerance, accretion of biomass, yield and quality of seeds, and plant adaptability [9]. The metabolism of nucleic acids and proteins and the gene expression is changed by BRs at the molecular level.

The extremely high activity of brassinosteroids has attracted the attention of many specialists in the field of analytical and synthetic chemistry, biochemistry, plant physiology, and agriculture. Recently, it has been reported that some natural BRs have (besides early reported, e.g., antibacterial, antiviral, antifungal, and neuroprotective) potent cell growth inhibitory activities in animal and human cancer cell lines without affecting the normal cell growth (BJ fibroblasts) [12, 13]. The presence of a lactone or ketone moiety in ring B and diol functions (2α, 3α- and 22*R*, 23*R*-) turned out to be an essential factor for high biological activity of such BRs (**Figure 1**).

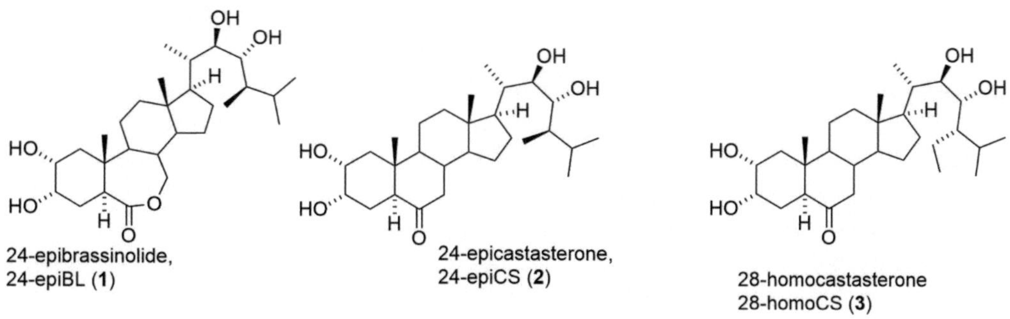

Figure 1. 24-epiBL (**1**) is the 24-(*R*)-epimer of the first isolated brassinosteroid, 24-epiCS (**2**), 28-homoCS (**3**).

This chapter is engaged in the recent developments in the synthesis of ^{3}H- and ^{14}C-radiolabeled analogs of the brassinolide.

2. Synthesis of tritium-labeled BRs

2.1. BRs with very high SA of tritium (~99 Ci/mmol)

For binding assays, study aimed at ligand-receptor-activity relationship is a high specific activity (SA), a bottom line requirement. The SA of such radio-labeled drugs need to be in scale of tenths of Ci/mmol. This critical precondition used to be an obstacle in the way of BR's studies for decades. The state-of-art strategy for such a labeling was reported by Marek et al. [14]. The methodology yields tritium-labeled BRs bearing a very high SA of 99.4 and 98 Ci/mmol (approx. 3.4 tritium enrichment per molecule), respectively. Convenient, a six-step synthetic sequence starting with the brassinosteroid *to be labeled* provides the desired tritium-multi-labeled product in sufficient yield (up to 40 mCi) with satisfactory radiochemical purity (>97%). The work is focused on the 24-[^{3}H]epibrassinolide [^{3}H]-**1** and 24-[^{3}H]epicastasterone [^{3}H]-**2**, both labeled in the side chain of BR skeleton on positions of carbon C-24, -25, -26, and -27 (**Figure 2**). The labeling strategy is designed to employ a radio-labeling step at the later stage of the synthetic sequence.

In 1998, Seto et al. described a fairly elegant strategy for the deuterium-multi-labeling of brassinolide in its side-chain [15]. The five-step reaction sequence was started by full protection of hydroxyl groups on the BR. The C-25 carbon was oxidized by freshly generated trifluoromethyldioxirane (TFD) that yielded appropriate hydroxy derivative. Its consecutive dehydration led to a mixture of $\Delta^{25(26)}$ and $\Delta^{24(25)}$ regiomers in the 65:35 ratio that was possible to separate after deprotection. The deuteration of $\Delta^{25(26)}$ regioisomer by deuterium gas catalyzed by Pd/C (1 atm, 25°C, 1 h) yielded [24, 25, 26, 27-^{2}H]brassinolide with 60% deuterium enrichment calculated from MS data. The ratio of the individual multi-deuterated species in the cluster was $^2H_2$:$^2H_3$:$^2H_4$:$^2H_6$:$^2H_7$ = 3:8:14:15:60. The basic idea of this methodology for usage at labeling with radioactive isotope tritium was waiting almost for two decades—then 24-[^{3}H]epiBL (**1**) and 24-[^{3}H]epiCS (**2**) was synthesized [14].

The protocol of Seto et al. paved the way for the synthesis of an unsaturated precursor for the intended synthesis of ^{3}H-labeled 24-epiCS [15, 16]. First, the 2,3-22,23-bisisopropylidene derivative **4** was prepared in a 96% yield by a reaction of 24-epiCS (**2**) with 2,2-dimethoxypropane catalyzed by *p*-toluenesulfonic acid [14]. Having 2,3,22,23-diisopropylidene-24-epicastasterone **4** available, the TFD hydroxylation of C-25 carbon with this particular derivative was studied. Unfortunately, the desired hydroxylation was accompanied by oxidation of alcohol C-3 affording appropriate hydroxyketone. Hence, more resistant 2,3,22,23-tetra-O-acetyl-24-epiCS was prepared. However, its C-25 hydroxylation did not proceed at all under the conditions used for the hydroxylation of **4**. Finally, 2,3-di-O-acetyl-22,23-isopropylidene-24-epicastasterone (**6**) was prepared which upon TFD hydroxylation at low temperature gave 2,3-di-O-acetyl-22,23-isopropylidene-25-hydroxy-24-epiCS (**7**) in 58% yield (**Figure 2**). The dehydration of **7** using thionyl chloride in pyridine at 0°C for 30 min afforded a mixture of prevailing

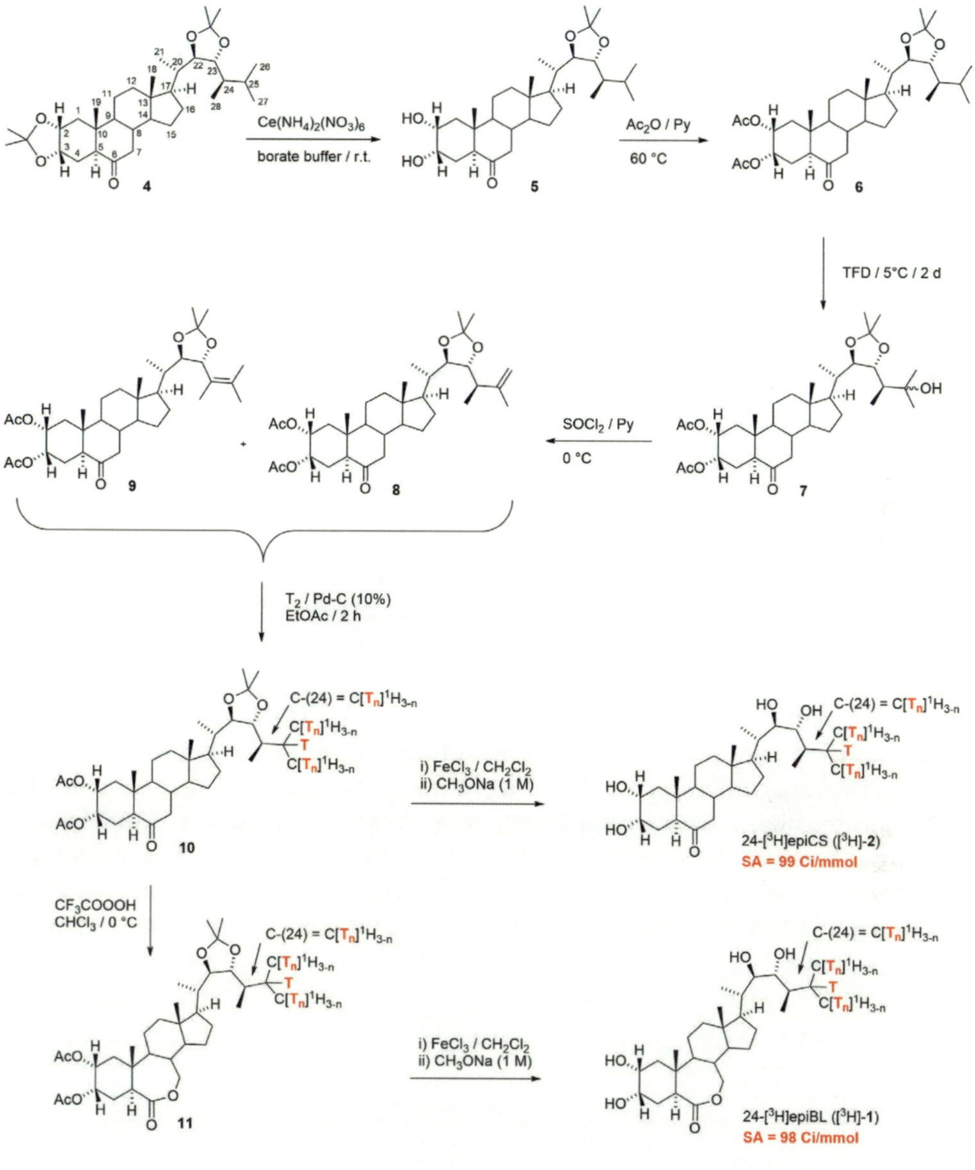

Figure 2. The successful approach for the synthesis of 24-[^{3}H]epiCS and 24-[^{3}H]epiBL with a high SA.

(22*R*,23*R*,24*R*)-2α,3α-diacetoxy-22,23-isopropylidenedioxy-24-methyl-5α-cholestan-25-ene-6-one (**8**) accompanied by its 24-ene regioisomer **9**. The separation of unsaturated regioisomers **8** and **9** from each other turned out to be infeasible using various conditions on high-performance liquid chromatography (HPLC). Importantly, it was possible to separate unsaturated derivatives **8** and **9** from 24-epicastasterone derivative **6** by HPLC. This fact eventually enabled the isolation of 261 mCi of (22*R*,23*R*,24*R*)-2α,3α-diacetoxy-22,23-isopropylidenedioxy-24-[24, 25, 26, 27-^{3}H]epicastasterone (**10**) after the catalytic tritiation of the mixture of the unsaturated derivatives **8** and **9** over Pd/C (10%) in ethyl acetate under carrier-free tritium gas (998 mbar) for 2 h. In one-pot synthesis, derivative **10** was deisopropylinated and deacetylated provided by HPLC purification, 40 mCi of 24-[24, 25, 26, 27-^{3}H]epicastasterone ([^{3}H]-**2**) with RCP >97% and SA_{MS} = 99.4 Ci/mmol [14].

To get [24, 25, 26, 27-^{3}H]epibrassinolide ([^{3}H]-**1**), the Baeyer-Villiger oxidation on fully protected 24-[24, 25, 26, 27-^{3}H]epicastasterone **11** was carried out by freshly prepared chloroform solution of trifluoroperoxyacetic acid [30% H_2O_2 (20 ml), trifluoroacetic acid (100 ml), $CHCl_3$ (1 mL)] cooled to 0°C by an ice bath [14] . The de-isopropylidation of the (22*R*,23*R*,24*R*)-2α,3α-diacetoxy-22,23-isopropylidenedioxy-24-[24, 25, 26, 27-^{3}H]epiBL (**11**) was carried out by wet $FeCl_3$, afterwards full deacetylation was accomplished by methanol solution of CH_3ONa. This conditions was made possible to obtain 3.5 mCi of pure 24-[24, 25, 26, 27-^{3}H]-epiBL ([^{3}H]-**1**) with RCP > 97% and SA_{MS} = 98 Ci/mmol. The ^{3}H NMR spectra of both [^{3}H]-**1** and [^{3}H]-**2** show tritium signals in C-25, C-26, and -27, and is in accordance with the determined SA, indicating 3.4 tritium atoms per molecule.

2.1.1. Stability of BRs possessing very high SA

The free BRs with a high SA are extremely sensitive to radiolysis if stored improperly. Authors reported one representative example—when a sample was evaporated to dryness and used for the NMR analysis (DMSO-d_6), the signals in ^{3}H NMR spectrum were difficult to assign [14]. Hence, the particular NMR sample was re-checked for purity by radio-HPLC, and indeed, only 12% of activity of desired 24-[24, 25, 26, 27-^{3}H]epiCS ([^{3}H]-**2**) was left whereas 88% of radioactivity was found in a broaden peak with higher retention on the column (**Figures 3** and **4**). Nevertheless, authors reported secure procedure to remove chromatographic solvents and formulate the high-SA BRs for application in biochemical experiments. Combined HPLC fractions (each about 2 mL) were first enriched with glycerol (300 mL) (acts as an antioxidant and also prevents risk of getting dryness of labeled samples) and 10 mg of (±)-6-hydroxy-2,5,7,8-tetramethylchroman-2-carboxylic acid (Trolox) (acts as additional antioxidant with no harm in eventual biological experiments) [14]. When methanol and water evaporated on

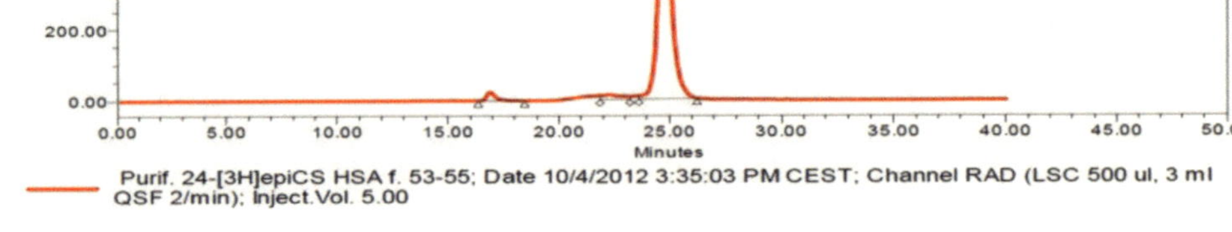

Figure 3. HPLC radiodetector chromatogram of [^{3}H]-**2** after purification.

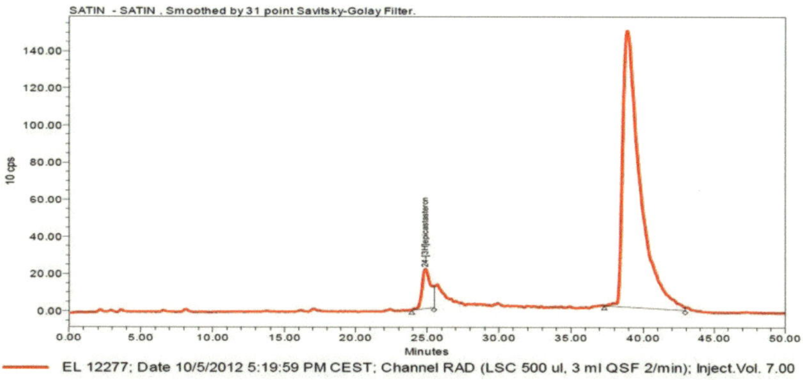

Figure 4. HPLC radiodetector chromatogram; fast decomposition of [3H]-2 when handled improperly - after its simple evaporation to dryness and storage in DMSO-d_6 overnight.

CentriVap, the residual glycerol solution was further diluted with water (2 mL), the radioactivity was determined and the concentration of 24-[24, 25, 26, 27-^{3}H]epiCS ([^{3}H]-**2**) was afterwards adjusted to 1 mCi/mL with a glycerol/water (1:1) mixture. The concentration of Trolox in the final formulation needed to be adjusted to 0.5%. For maximal stability of prepared samples, storage of 1 mCi aliquots in liquid nitrogen is recommended.

2.2. BRs with reasonable high SA of tritium

2.2.1. Reductive tritium-dehalogenation of generated chlorocarbonates (6 Ci/mmol)

To get polyhydroxylated steroid regio- and enantio-specifically labeled on the un-exchangeable position of C-3, a general procedure can be effectively used (**Figure 5**) [17]. A suitable precursor for the introduction of tritium, 3β-chloro-2,3-carbonate derivative, is synthetically affordable by a short-reaction sequence from a 2α,3α-dihydroxy steroid *to be labeled* (**Figure 6**) [18]. Chlorocarbonate undergoes reductive tritium dechlorination catalyzed by the [Pd0]/Et_3N system, providing 28-[3β-^{3}H]homoCS, 24-[3β-^{3}H]epiCS, and 24-[3β-^{3}H]epiBL, respectively, in good yield and with high SA (5.8 Ci/mmol; 0.2 tritium per molecule) (**Figure 5**) [19]. A crucial aspect in the reductive dehalogenation of the chloro derivative is the choice of a solvent that would provide a reasonable yield [17]. The optimized reaction conditions turned out to be PdO/$CaCO_3$ (5%)/Et_3N/chlorocarbonate (2:6:1) dissolved in dry EtOAc for all investigated steroids. The successful ^{3}H-labeling experiments have proven the stereo selectivity of the reductive dehalogenation and afforded a product with high specific activity (5.8 Ci/mmol).

The synthetic procedure starts with transformation of the vicinal 2α,3α-diols of appropriate BR to α-hydroxy ketone by oxidation with a freshly generated dimethyldioxirane (DMD). Such α-hydroxy ketone moiety proved to be an excellent substrate for high-yield enantiospecific formation of 3β-chloro-2,3-carbonate by a reaction with easy-to-handle triphosgene. The key substrate for reductive dechlorination—3β-chloro-2,3-carbonate—was synthesized by a three-step reaction sequence in an overall yield of 46–55% (**Figure 6**; representative synthesis of 24-[^{3}H]epiCS). To improve the solubility of the starting steroid *(to be labeled)* in a

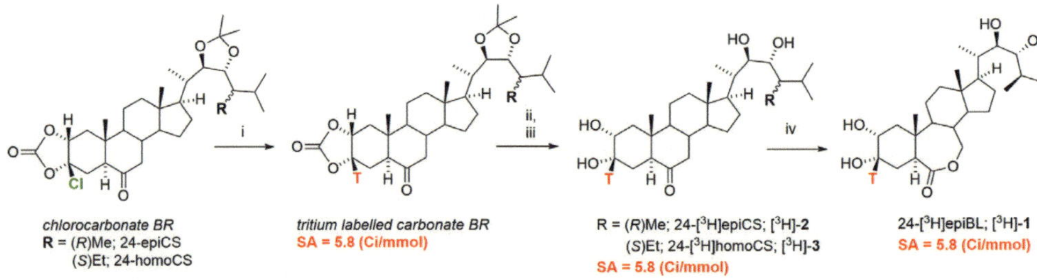

Figure 5. Tritium Pd-catalyzed reductive dehalogenation; (i) T_2/PdO/$CaCO_3$/Et_3N; (ii) Fe(III), CH_2Cl_2; (iii) NaOH, 1,4-dioxane; (iv) H_2O_2/TFA, 0°C, 30 min, r.t., 4 h, $CHCl_3$.

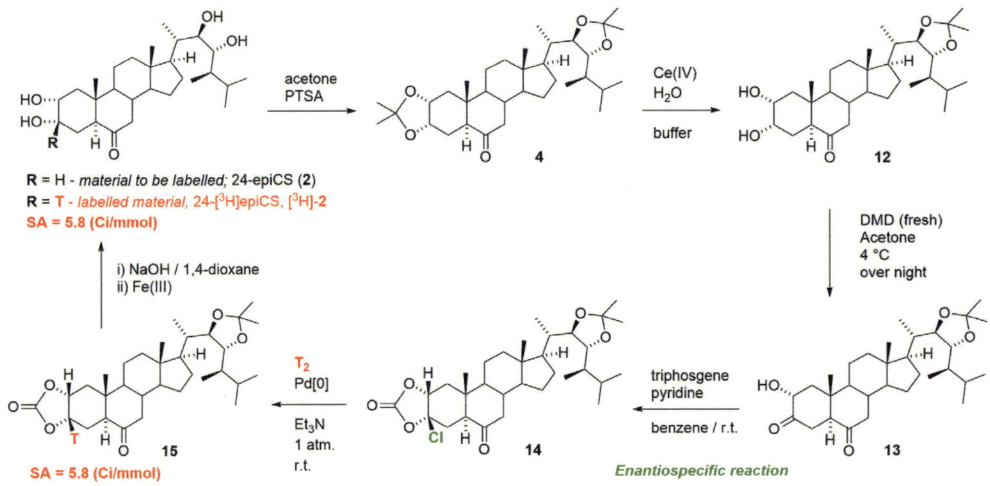

Figure 6. Reaction sequence of the synthesis of 24-[^{3}H]epiCS; [^{3}H]-**2**.

non-polar solvent, 2,3- and 22,23-vicinal diols were protected. The isopropylidation of both vicinal diol groups by 2,2-dimethoxypropane (10 eqv.) catalyzed by *p*-toluenesulphonic acid in dry CH_2Cl_2 turned out to be an elegant protecting strategy. Full conversion was reached in 30 h and the pure product **4** was isolated. Acetonide **4** was further oxidized at the position of C(3)-OH by freshly synthesized dimethyldioxirane (DMD) [20]. The reaction carried out in CH_2Cl_2 overnight at 4°C in dark afforded the desired α-hydroxy ketone **13** accompanied by its regioisomer 3-hydroxy ketone, in a ratio of 10:1 in favor of isomer **13**. In general, α-hydroxy ketone **13** is supposed to be isolated in a higher yield when the oxidation of 2,3-unprotected diols takes place; a selective deprotection of acetonide by $Ce(NH_4)_2(NO_3)_6$ in borate buffer followed by the DMD oxidation of the vicinal 2,3-diol group used to provide a higher overall yield of about 5% [21]. However, in the case of the 28-HCS derivative, a partial deprotection-oxidation reaction sequence afforded a drop of an isolated yield of about 15% compared to direct oxidation of the protected derivative. The 3β-chloro-2,3-carbonate **14** was synthesized by a stereospecific reaction of **13** with triphosgene in dry benzene providing a quantitative yield in 3 h (**Figure 7**).

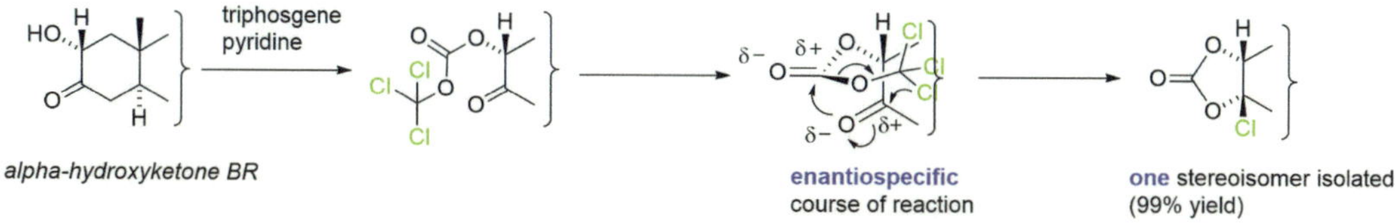

Figure 7. Mechanism of enantiospecific formation of 3β-chloro-2,3-carbonate derivatives of BR.

The catalytic reductive dehalogenation of BR–chlorocarbonates was studied with deuterium in the system of $^2H_2$/Pd[0]/Et_3N providing appropriate 3β-deutero-2,3-carbonates with 70–80% deuterium enrichment (based on ^1H NMR) at the C-3 position and with an isolated yield of up to 65% (initially on a cheap pregnane analog available in multigame scale). The best results for deuterium dehalogenation were achieved with the molar ratio of PdO/$CaCO_3$(5%)/Et_3N/chlorocarbonate being 2:6:1 for every BR derivative at a short period of time (6 h). When the amount of the base was too high, it diminished the yield of [^{2}H]-labeled ethylene carbonate. Various catalysts such as Pd/C (either 5 or 30%), PdO/$BaSO_4$ (10%) used for reduction yielded lower yield of desired carbonate (15–19%). Authors disclosed very significant solvent effect with an impact on the isolated yield as well as the by-products formation. Briefly, the best results provided EtOAc (dry), giving up to a 65% yield of labeled carbonate with 80% ^{2}H-enrichment. Other solvents used for reduction provided both low conversion and isolated yield of carbonate (0–19%). Androstane chlorocarbonate employed for reductive dehalogenation under similar condition as for pregnane analog provided analogous results (58% yield, 75% ^{2}H-enrichment at C-3). Surprisingly, in addition to desired labeled carbonate two by-products were detected, isolated, and afterwards characterized in that experiment—the multi-labeled ketone and the multi-labeled alcohol, both in the yield of 15%. The reaction course toward formation of both by-products was further accelerated while protic solvent (MeOH) was used in the reaction [43% (ketone) and 40% (alcohol)]. The reaction conditions used for the labeling of 24-epiCS were the same as described above for the other two steroids. A full conversion of appropriate chlorocarbonate was obtained after a 6-h reaction [PdO/$CaCO_3$ (5%)/Et_3N/substrate 2:6:1] in dry EtOAc. The isolated yield of labeled carbonate was determined to be 31% [D/H at C-3 = 70:30]. Both the by-products, ketone (20%) and alcohol (13%), were isolated too. The use of DMF as solvent reduced the conversion of chlorocarbonate to 45%, the yield of labeled carbonate down to 19% and the formation of by-products to 11 and 10%, respectively.

As was already mentioned, traces of water (partly synthesized by the reduction of PdO with D_2 gas) play a crucial role in the suggested mechanism of by-product formation (**Figure 8**) [17]. The initial oxidative addition of *in situ* generated Pd[0] into the chlorocarbonate C(3)-Cl bond forms an organopalladium compound **16**. Such palladium could be partly substituted by traces of water to form appropriate hydroxyl carbonate **17**, which afterwards undergoes ring opening, leading to ketone **18**. Unsaturated ketone **19** is formed by an elimination of carbonic acid from ketone **18** (Pathway A), which could be *in situ* reduced by a system of D_2/Pd[0] providing 1,2-deuterium-labeled ketone **20** and **21**, respectively. The other discussed pathway (B) involves an elimination of carbon dioxide, affording α-hydroxyketone, and the subsequent

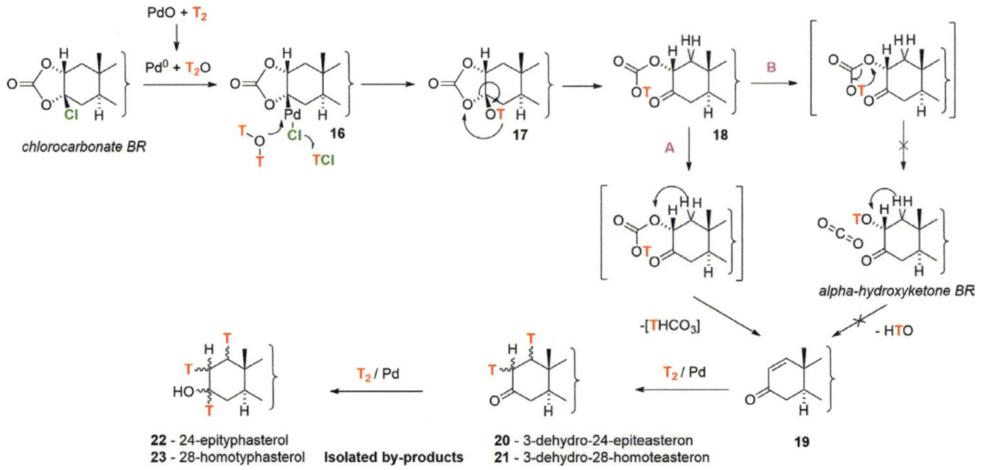

Figure 8. Mechanism of multi-labeled by-products formation.

elimination of H_2O leading to **19**. Authors declare a statement that theoretical pathway B was not supported by any analytical findings. First, the α-hydroxyketone was not detected in the reaction mixture. Moreover, the single experiment using synthesized α-hydroxyketone and reaction conditions simulating conditions used for reductive deuterium dehalogenation, yielded ultimately 2α,3β-dihydroxy-3α[^{2}H] derivative in an isolated 85% yield and no other product was identified in the reaction mixture. In advance, the absolute configuration at C-2 and C-3 of 2α,3β-dihydroxy derivative was confirmed by was confirmed by X-ray analysis affording ORTEP diagram. A regular ESI-MS analysis of **20** and **21** measured in CH_3OH has confirmed the structure of multi-labeled ketones with the distribution of deuterium 1D/2D being 70:30. The second by-product, multi-labeled alcohols **22** and **23**, are most likely formed by a Pd-catalyzed reduction of the ketones **20** and **21** by D_2. The structure of **22** and **23** was confirmed by ^{1}H NMR as well as ESI-MS with a deuterium distribution 1D/2D/3D being 10:50:40.

In view of the favorable results of deuterium experiments, this protocol was followed using tritium gas. Tritium dehalogenation experiment was designed following the optimized reaction conditions [PdO/$CaCO_3$ (5%)/Et_3N/substrate 2:6:1, dry EtOAc]. ^{3}H-labeled 24-epiCS and 24-epiBL, was synthesized when the appropriate 3β-chloro-2,3-carbonate **14** was used for a reaction with carrier-free tritium gas (600 mbar), PdO/$CaCO_3$ (5%), and in presence of base Et_3N. The reaction proceeded at 25°C for 17 h (**Figure 6**). The reduction yielded 5.9 mCi of the desired ^{3}H-labeled carbonate **15**, and two further unidentified by-products (21.3 and 12.4 mCi) were detected. By ^{1}H NMR was determined a specific activity of **15** at 5.8 Ci/mmol (which accounts for 0.2 tritium atom per molecule). Compared to previous deuterium comprehensive study, the obtained tritium enrichment was lower than expected. Authors speculate that the significant drop in SA was caused by the reduced pressure of tritium gas (600 mbar $^3H_2$) compared to deuterium gas (950 mbar of $^2H_2$) used for the reduction. The only signal in the ^{3}H NMR spectrum (the singlet at δ 4.8 ppm) explicitly determined the regio- and stereo-specificity of the reduction. The deprotection of the isopropylidene group in the side chain was carried out by wet $FeCl_3$ in CH_2Cl_2. The last step—hydrolysis of ^{3}H-labeled carbonate **15**—was accomplished by NaOH (0.5 M) in 1,4-dioxane (1:1). Desired [^{3}H]-**2** (3.8 mCi, 5.8 Ci/mmol) was purified by preparative radio-HPLC. Aliquot of [^{3}H]-**2** was further employed for Baeyer-Villiger oxidation leading to [^{3}H]-**1** (0.3 mCi, SA 5.8 Ci/mmol).

The SA of isolated [^{3}H]-**2** was about four times lower than was predicted based on the previous comprehensive deuterium-using study. To further investigate the influence of a metal catalyst on the SA and the yield of the labeled product, it was considered the use of the Pd[0] catalyst instead of the Pd[II]O catalyst, where the generation of Pd[0] is accompanied by the formation of tritiated water. The suggested mechanism of the by-products formation explained how the traces of water significantly reduce the yield of the labeled product desired (**Figure 8**). Hence, to suppress synthesis of multi-labeled by-products Pd on charcoal was employed at synthesis of 28-[3β-^{3}H]homocastasterone [19]. The catalyst/base/substrate ratio for the tritium experiment was kept identical to the previous experiment of the tritium labeling of 24-epiCS—2:6:1. Carrier-free tritium released over the reaction mixture (738 mbar, 11.5 Ci, 180 μmol) was left to react for 24 h at room temperature. Then, the analytical radio-HPLC proved the formation of the desired ^{3}H-labeled carbonate (13.3 mCi) and two other by-products (42.8 and 24.7 mCi) were detected. The SA of ^{3}H-carbonate was determined by ^{1}H NMR as 5.8 Ci/mmol (based on the decrease of the corresponding ^{1}H signal intensity in the labeled position). The deprotection of the 22,23-isopropylidene group by treatment with wet $FeCl_3$ in CH_2Cl_2 was completed within 10 min. The crude 3β-tritio-2,3-carbonate was directly hydrolyzed by 0.5 M aqueous NaOH in 1,4-dioxane. The crude product was purified on the semi-prep radio-HPLC, affording 5.3 mCi of [^{3}H]-(**3**) of SA 5.8 Ci/mmol with radiochemical purity (RCP) >97%. Both by-products were isolated, and ^{3}H and ^{1}H NMR measured. The structure of multi-labeled by-products is believed to be similar to those that were recently described (**Figure 9**) for androstane and 24-epiCS (i.e., labeled 24-epityphasterol and 3-dehydro-24-epiteasteron, respectively) derivatives. In this case, these by-products are supposed to be 22,23-isopropylidene-protected multi-labeled [1, 2, 3-^{3}H]-28-homotyphasterol and [1, 2-^{3}H]-3-dehydro-28-homoteasterone derivatives (**Figure 9**).

2.2.2. *Catalytic reduction of 24-methylene BRs (SA = 2 Ci/mmol)*

An elegant and fast strategy to get BRs labeled by tritium was briefly communicated by Yokota et al. [22]. [^{3}H]-BRs synthesized on demand at Amersham International (Amersham, UK) were used for comparative analysis in stems and seeds by radioimmunoassay. The labeling strategy was based on platinum-catalyzed reduction of 24-methylene position of available BR (dolichosterone and dolicholide) in carrier-free tritium atmosphere (**Figure 10**). Reduction of dolichosterone afforded two epimers 24-[24, 28-^{3}H]castasterone (8.1 mCi, SA = 2.2 Ci/mmol) and 24-[24, 28-^{3}H]epicastasterone [^{3}H]-(**2**) (6.5 mCi, SA not determined), respectively, which were then separated on HPLC. Analogically, reduction of dolicholide provided 24-[24, 28-^{3}H] brassinolide (10.6 mCi, SA = 2.3 Ci/mmol) and 24-[24, 28-^{3}H]epibrassinolide [^{3}H]-(**1**) (4.0 mCi, SA not determined). Authors stated unexpectedly low SA of gained products (theoretical SA was supposed to be over 40 Ci/mmol). Detailed synthetic procedure and analysis were not reported.

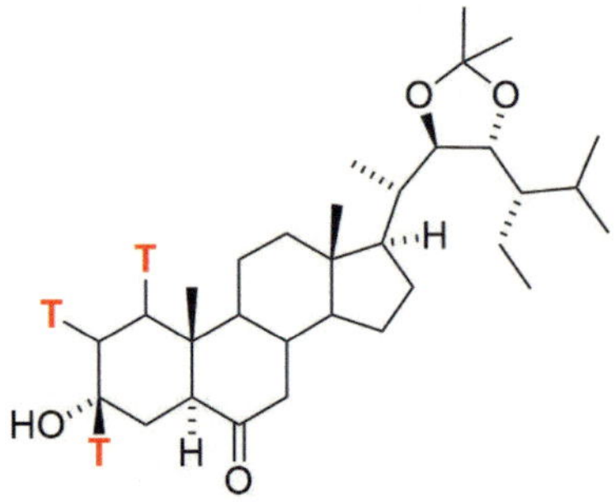

22,23-(isopropylidenedioxy)-
[1,2,3-^{3}H]-28-homotyphasterol (**23**)

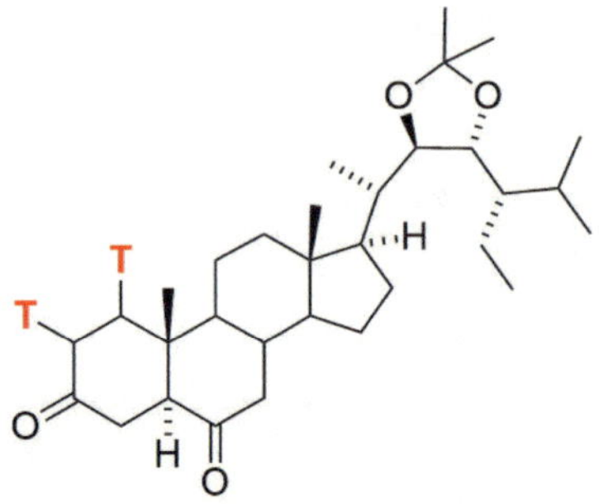

22,23-(isopropylidenedioxy)-
[1,2-^{3}H]-3-dehydro-28-homoteasterone (**21**)

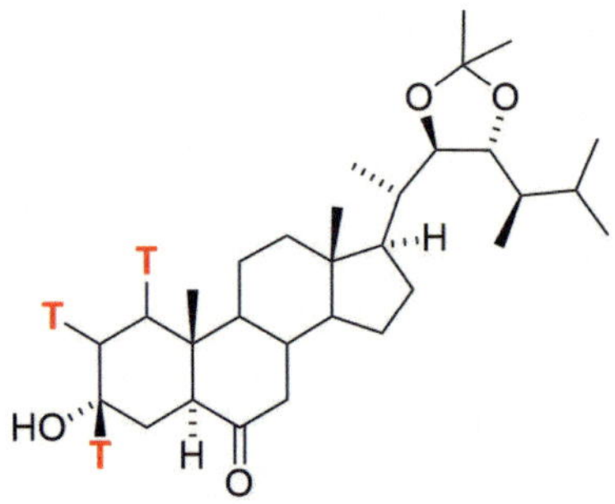

22,23-(isopropylidenedioxy)-
[1,2,3-^{3}H]-24-epityphasterol (**22**)

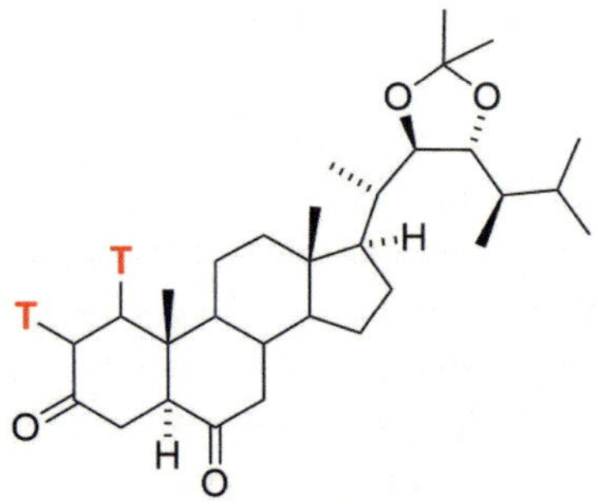

22,23-(isopropylidenedioxy)-
[1,2-^{3}H]-3-dehydro-24-epiteasterone (**20**)

Figure 9. Multi-labeled by-products.

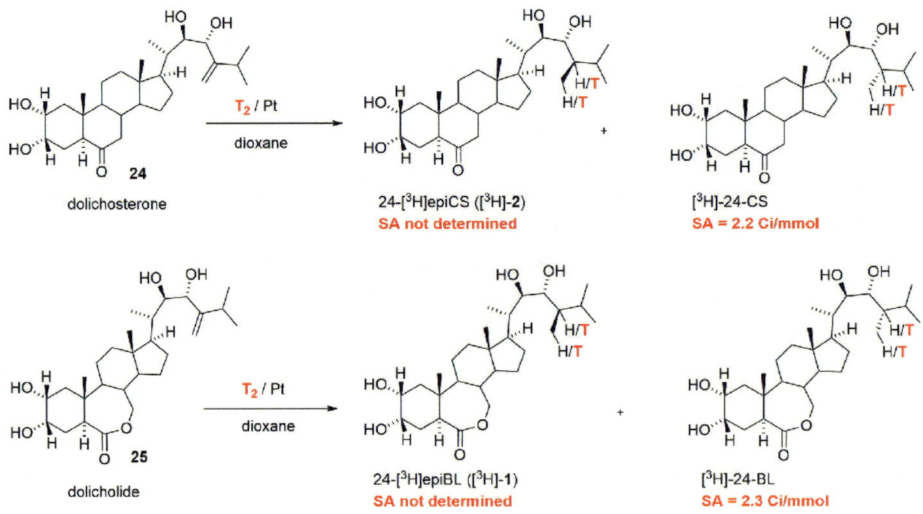

Figure 10. Metal-catalyzed 24-methylene BR tritiation.

2.3. BRs with very low SA of tritium (10^{-3} Ci/mmol)

Tritium-labeled protected 24-epiCS and 24-epiBL [^{3}H]-**1** were for the first time prepared by Kolbe et al. [23]. By very simple procedure of base-catalyzed exchange reaction with tritiated water (SA = 20.10^{-3} Ci/mmol—accounts for 0.7 × 10^{-3} tritium atom per molecule HTO) to the enolizable alpha positions of C-6 ketone of tetracetate 24-epiCS **27** (**Figure 11**). By this method, it is theoretically possible to exchange hydrogen on three distinctive positions (C-5α, C-7α, and C-7β). However, the obtained SA of BRs was very low (6.10^{-3} Ci/mmol), about 3-4 orders lower than that provided by methods described above. Because of low SA of such prepared material cannot be used in receptor studies at all. The other drawback of this procedure is chemical exchangeability of labels, which inevitably leads to loss of label in protic solvents (water) during biological experiments and disqualifies this approach from use in ADME studies. Three-step reaction sequence provided 24-[5, 7, 7-^{3}H]epiBL with SA 6.10^{-3} Ci/mmol; labeling step leading to protected 24-[^{3}H]epiCS **27**, followed by a Baeyer-Villiger oxidation with CF_3COOOH and sequence was accomplished by hydroxy group deprotection [20] . The same sequence was also performed with 2,3,22,23-bis-isopropylidenedioxy-24-epiCS **29** as starting compound. The advantage of the use of isopropylidene protecting groups over acetate group is the deprotection step. The lactone ring is stable under acidic hydrolysis conditions used for isopropylidene deprotection, which occurs simultaneously with Baeyer-Villiger oxidation. The basic hydrolysis needed for the acetate group cleavage leads to lactone hydrolysis and acid catalyzed re-lactonization is needed (**Figure 11**).

A base-catalyzed exchange was used for labeling of biogenetic brassinosteroids precursors [24]. 24-[5, 7, 7-^{3}H]epiteasterone (SA = 1.5 10^{-3} Ci/mmol), 6-oxo-24β-methyl-22-dehydro[5, 7, 7-^{3}H]cholestenol (SA = not indicated), and 6-oxo-24-[5, 7, 7-^{3}H]epicampestanol (SA = 3.5 10^{-3} Ci/mmol), respectively, were partly labeled on positions of C-5 and C-7 by reaction in sealed

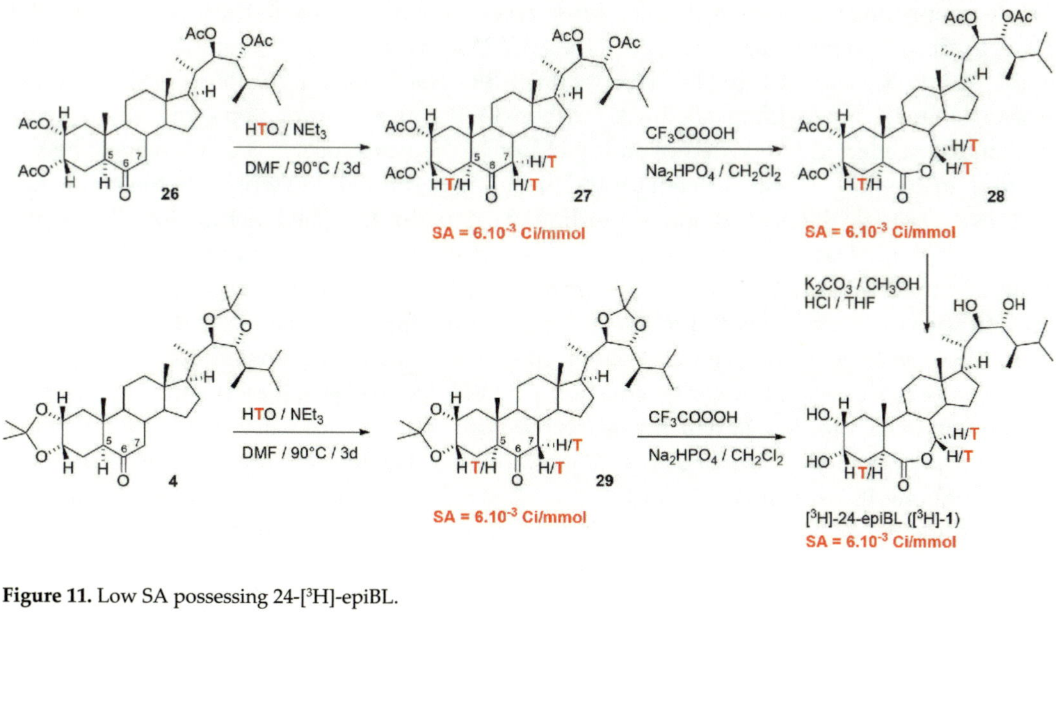

Figure 11. Low SA possessing 24-[^{3}H]-epiBL.

ampoule with tritiated water (SA = 14.10^{-3} Ci/mmol—accounts for 0.5 × 10^{-3} tritium atom per molecule HTO) in presence of base Et_3N (**Figure 12**). As mentioned above this simple methodology affords poor SA (in order of 10^{-3} Ci/mmol) of BRs labeled on an exchangeable positions thus prone to loss of label later on. On the other hand, this approach provides high yield (>45%) of desired material.

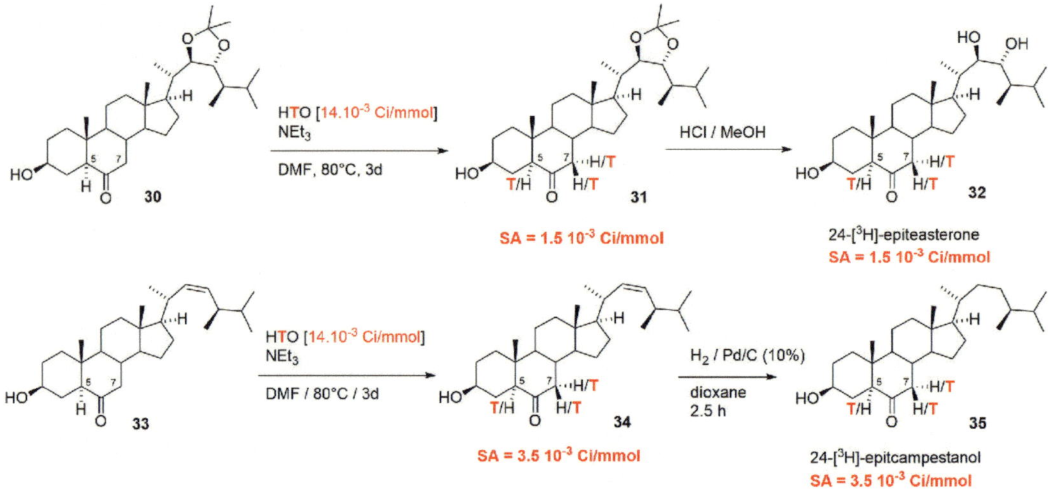

Figure 12. 24-[^{3}H]epiteasterone and 6-oxo-24-[^{3}H]epicampestanol.

2.4. BRs labeled by carbon-14 (SA = 56.8 × 10^{-3} Ci/mmol)

Till now, only one report on ^{14}C-labeled brassinosteroids is available in the literature. In 1989, Seo et al. described the synthesis of [^{14}C]-labeled (22*R*,23*R*)- and (22*S*,23*S*)-epiCS **50** and **51**, and epiBL **52** and **53**, respectively [25]. The obtained BRs possessed top SA (56.8 × 10^{-3} Ci/mmol) possible to reach while doing labeling by carbon-14. Because of multi-step synthetic sequence (>10 steps) needed, the reported overall yield of labeled products is 3.2 and 4.5%. The C-4 position in 24-epiBL **52** was selected for [^{14}C]-labeling because of its stability to metabolic loss and easy way to do the preparation. According to the well-established method for ^{14}C incorporation into the C-4 position of steroids, the enol lactone **41** was synthesized from the starting material brassicasterol **36** in five steps (**Figure 13**). Bridged ketone **42** was prepared from lactone **41** after its treatment with [^{14}C]methyl iodide. Alkaline treatment of **42** provided [4-^{14}C] enone **43**. Gentle acid catalyzed acetylation of enone **43** with isopropenyl acetate led to the enol acetate **44** that was afterwards reduced by $NaBH_4$ to give [4-^{14}C]brassicasterol **45**. Simple mesylation of **45** afforded **46** which was treated with sodium carbonate providing the major product 3,5-cyclo-6-ol **47** isolated in 92% yield (**Figure 14**). Jones oxidation of **47** then gave 3,5-cyclo-6-one **48**. Rearrangement of **48** moderated by lithium bromide and camphor sulfonic acid in dimethyl acetamide afforded 2,22-diene-6-one **49** in quantitative yield. Oxidation of

49 with osmium tetroxide led to a stereo isomeric mixture of 2,3,22,23-tetraols–24-(22*R*,23*R*)-[^{14}C]epiCS **50** and 24-(22*S*,23*S*)-[^{14}C]epiCS **51** which were separated. The final Baeyer-Villiger oxidation of ketone C-6 with TFA in dichloromethane afforded the (22*R*,23*R*)-7-oxa-lactone **52** accompanied with its 6-oxa isomer **56** as a minor product. Analogically, (22*S*,23*S*)-tetraol provided the (22*S*,23*S*)-7-oxa-lactone **53** accompanied with its 6-oxa isomer **57**. The ultimate products (22*R*,23*R*)-24-[^{14}C]epiBL **52** and (22*S*,23*S*)-24-[^{14}C]epiBL **53** were isolated in overall 3.20 and 4.46% radiochemical yield (toward $Ba^{14}CO_3$ used as an initial source or radio-label), respectively. SA of prepared BRs was 56.8 × 10^{-3} Ci/mmol.

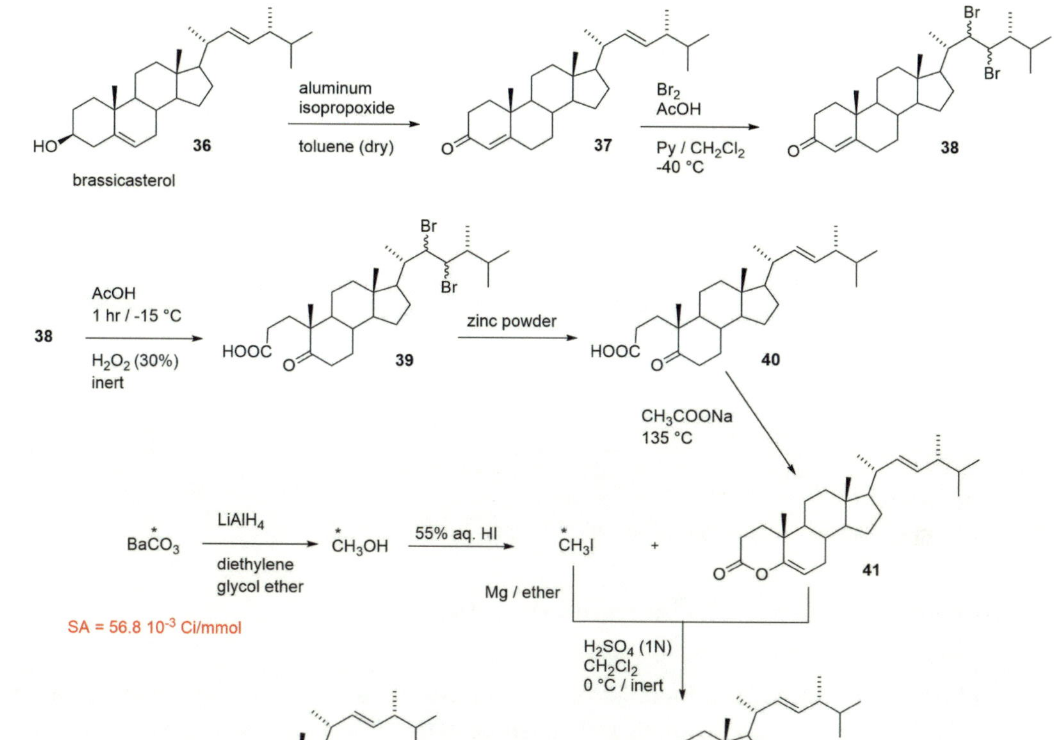

Figure 13. Synthetic pathway to 24-[^{14}C]epiCS and 24-[^{14}C]epiBL.

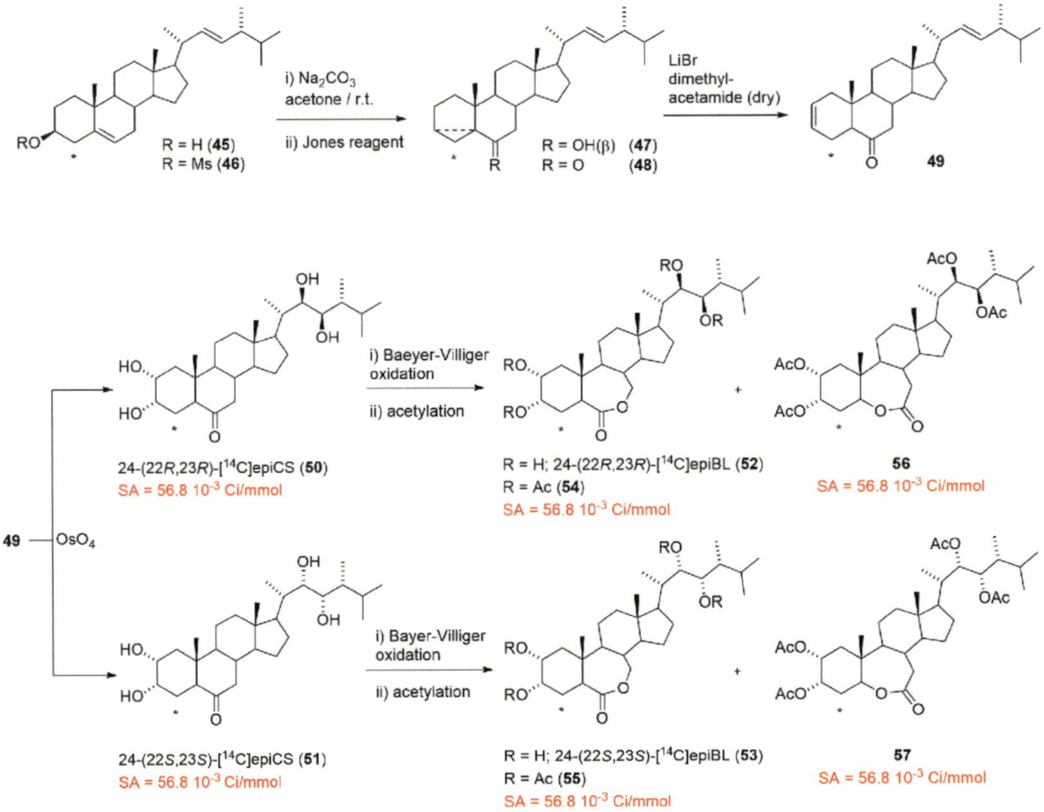

Figure 14. Synthetic pathway to 24-[14C]epiCS and 24-[14C]epiBL.

3. Equipment of the radio-isotope laboratory (IOCB)

The author of this chapter is a member of the Radio-isotope laboratory, a service group of the IOCB CAS, working as a synthetic radiochemist in the production of radioactive molecular tracers. The laboratory is classified for handling of the open sources of ionizing radiation in quantities authorized for laboratories of II category according to Czech bylaw 307/2002 Sb for research and development and educational purposes. It is currently authorized to work with the main radioactive isotopes used in research, for example, ^{3}H, ^{14}C, ^{32}P, ^{33}P, ^{35}S, ^{51}Cr, ^{54}Mn, ^{55}Fe, ^{99}Tc and ^{125}I. The main purpose of the laboratory is to provide a series of highly specific facilities, equipment and services fully adapted to researchers' needs and maintained in optimum conditions, always in line with applicable legislation to ensure that all personnel is fully protected and ensuring the physical safety of the materials used and the environment. The laboratory has an equipment, stuff, and knowledge to cooperate with other chemists and biologists to provide them with custom synthesis of radio-labeled compounds, especially commercially unavailable compounds. Following time-depending stability of synthesized radio-labeled compounds is one of our basic services provided to biologists.

http://dx.doi.org/10.5772/intechopen.68584

The key instrument of the laboratory is a glove box with tritiation manifold from RC-TRITEC AG (Teufen, Switzerland) suitable for handling 100–1000 Ci of carrier-free tritium gas (**Figure 15**). Tritiation manifold is based on U-Bed Technology to provide fresh,

Figure 15. The equipment for safe handling of tritium gas; the tritium manifold (RC-TRITEC AG) placed in the glove box; a scrubber for decontamination placed behind the glove box.

^{3}He-free tritium for tritiation by simply heating the UT_3-bed, also allowing the recovery of surplus gas after completion of a reaction (**Figure 16**). During the operation of the manifold, the internal atmosphere of the glove box is continually decontaminated by a scrubber equipped with catalytic oxidation of gaseous tritium to tritiated water, which is trapped

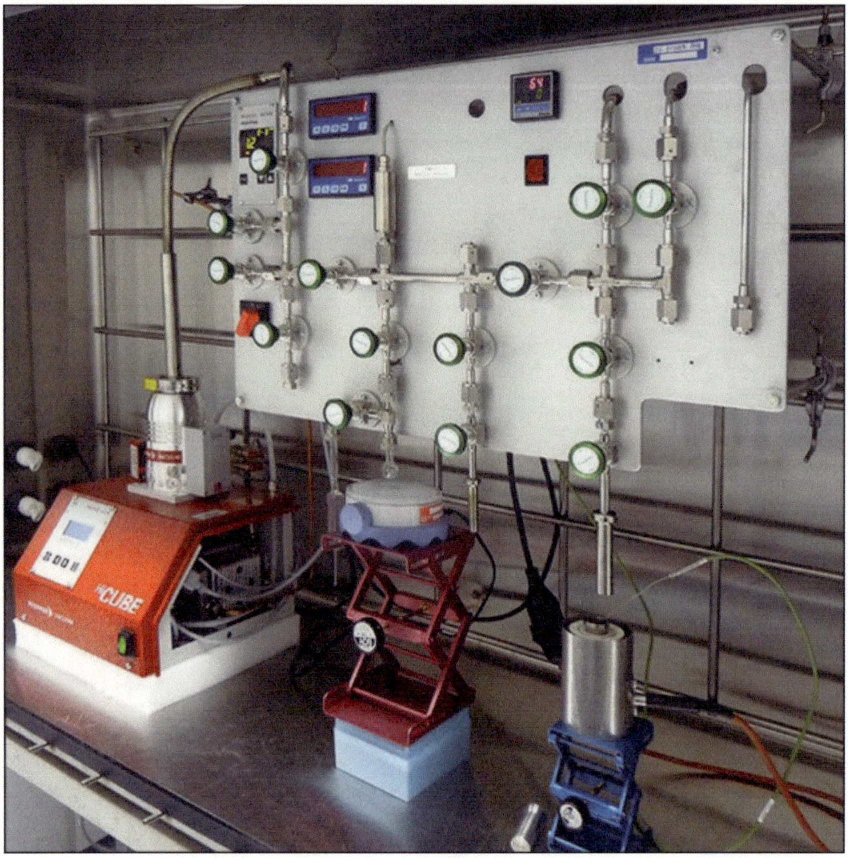

Figure 16. The equipment for safe handling of tritium gas; the tritium manifold (RC-TRITEC AG) placed in the glove box.

on a molecular sieve. ^{3}H NMR measurements were performed on a Bruker Avance II 300 MHz in the laboratory. Liquid scintillation analyser Tri-Carb 2900TR (Perkin Elmer) was used for detecting small amounts of α, β, and γ radioactivity. Mobile contamination monitor CoMo 170 (GRAETZ Strahlungsmeßtechnik GmbH) was used for the high-sensitive and nuclide referred measurement of surfaces with regard to α-, β-, and γ-contaminations when handling ionizing radiation. Analytical -preparative radio-HPLC (pump Waters 600, UV detector Waters 2487, radio chromatogram detector Ramona with analytical cell (LSC) and solid scintillator preparative cell (Raytest, Germany), data management software Empower 2 from Waters). Basic radiation protection equipment and waste disposal management.

Acknowledgements

The authors thank the Czech Academy of Sciences for the financial support within the program RVO: 61388963.

Author details

Aleš Marek

Address all correspondence to: ales.marek@uochb.cas.cz

Radioisotope Laboratory, Institute of Organic Chemistry and Biochemistry of the Czech Academy of Sciences, Czech Republic, Flemingovo náměstí, Praha

References

[1] Marathe PH, Shyu WC, Humphreys WG. The use of radiolabeled compounds for ADME studies in discovery and exploratory development. Current Pharmaceutical Design. 2004;**10**(24):2991-3008

[2] Penner N, Klunk LJ, Prakash C. Human radiolabeled mass balance studies: Objectives, utilities and limitations. Biopharmaceutics & Drug Disposition. 2009;**30**(4):185-203

[3] Voges R, Heys JR, Moenius T. Preparation of The Atrium, Southern Gate, Chichester, West Sussex, United Kingdom. Compounds Labeled with Tritium and Carbon-14. John Wiley & Sons; 2009. pp. 1-21

[4] Krauser JA. A perspective on tritium versus carbon-14: Ensuring optimal label selection in pharmaceutical research and development. Journal of Labelled Compounds and Radiopharmaceuticals. 2013;**56**(9-10):441-446

[5] Villa-Alfageme M, Manjon G. Low-level measurements of tritium in water. Applied Radiation and Isotopes. 2004;**61**(2-3):319-323

[6] L'Annunziata MF. Handbook of Radioactivity Analysis. Academic Press; The Boulevard, Langford Lane, Kidlington, UK. 2012. pp. 424-622

[7] Oklestkova J, Rarova L, Kvasnica M, Strnad M. Brassinosteroids: Synthesis and biological activities. Phytochemistry Reviews. 2015;**14**(6):1053-1072

[8] Grove MD, Spencer GF, Rohwedder WK, Mandava N, Worley JF, Warthen JD, et al. Brassinolide, a plant growth-promoting steroid isolated from Brassica-Napus pollen. Nature. 1979;**281**(5728):216-217

[9] Bajguz A. Brassinosteroids—Springer Science+Business Media B.V. Van Godewijckstraat Dordrecht Netherlands Occurence and chemical structures in plants. In: Hayat S, Ahmad A, editors. 2011. pp. 1-27.

[10] Bajguz A, Tretyn A. The chemical characteristic and distribution of brassinosteroids in plants. Phytochemistry. 2003;**62**(7):1027-1046

[11] Bajguz A. Metabolism of brassinosteroids in plants. Plant Physiology and Biochemistry. 2007;**45**(2):95-107

[12] Malikova J, Swaczynova J, Kolar Z, Strnad M. Anticancer and antiproliferative activity of natural brassino steroids. Phytochemistry. 2008;**69**(2):418-426

[13] Steigerova J, Rarova L, Oklest'kova J, Krizova K, Levkova M, Svachova M, et al. Mechanisms of natural brassinosteroid-induced apoptosis of prostate cancer cells. Food and Chemical Toxicology. 2012;**50**(11):4068-4076

[14] Marek A, Patil MR, Elbert T. The labeling of brassinosteroids by tritium. RSC Advances. 2015;**5**(80):65214-65220

[15] Seto H, Fujioka S, Koshino H, Yoshida S, Watanabe T, Takatsuto S. A general approach to synthesis of labeled brassinosteroids: Preparation of 25,26,27-H-2(7) brassinolide with 60% isotopic purity from the parent brassinolide. Tetrahedron Letters. 1998;**39**(41): 7525-7528

[16] Seto H, Fujioka S, Koshino H, Yoshida S, Tsubuki M, Honda T. Synthesis and biological evaluation of extra-hydroxylated brassinolide analogs. Tetrahedron. 1999;**55**(28):8341-8352

[17] Marek A, Klepetarova B, Elbert T. A facile method for steroid labeling by heavy isotopes of hydrogen. Tetrahedron. 2015;**71**(29):4874-4882

[18] Marek A, Patil MR, Klepetarova B, Kohout L, Elbert T. A stereospecific pathway for the introduction of deuterium on the brassinosteroid skeleton by reductive dechlorination of chlorocarbonates. Tetrahedron Letters. 2012;**53**(16):2048-2050

[19] Elbert T, Patil MR, Marek A. Enantiospecific tritium labeling of 28-homocastasterone. J Label Compd Radiopharm. 2017;**60**(3):176-82. DOI:10.1002/jlcr.3488

[20] Marek A, Klepetarova B, Elbert T. The introduction of a double bond on the steroid skeleton—The preparation of enol silyl ether derivatives from vicinal diols. Collection of Czechoslovak Chemical Communications. 2011;**76**(5):443-456

[21] Marek A, Klepetarova B, Elbert T. A study of the reactivity of polyhydroxylated sterol derivatives. Asian Journal of Organic Chemistry. 2015;**4**(8):808-817

[22] Yokota T, Watanabe S, Ogino Y, Yamaguchi I, Takahashi N. Radioimmunoassay for brassinosteroids and its use for comparative analysis of brassinosteroids in stems and seeds of Phaseolus-Vulgaris. Journal of Plant Growth Regulation. 1990;**9**(3):151-159

[23] Kolbe A, Marquardt V, Adam G. Synthesis of tritium labeled 24-epibrassinolide. Journal of Labelled Compounds and Radiopharmaceuticals. 1992;**31**(10):801-805

[24] Kolbe A, Schneider B, Voigt B, Adam G. Labelling of biogenetic brassinosteroid precursors. Journal of Labelled Compounds and Radiopharmaceuticals. 1998;**41**(2):131-137

[25] Seo S, Nagasaki T, Katsuyama Y, Matsubara F, Sakata T, Yoshioka M, et al. Synthesis of (22R,23R) and (22S,23S)-4-C-14-24-epibrassinolide. Journal of Labelled Compounds and Radiopharmaceuticals. 1989;**27**(12):1383-1393

Chapter 3

Salicylic Acid: An All-Rounder in Regulating Abiotic Stress Responses in Plants

Mirza Hasanuzzaman, Kamrun Nahar,
Tasnim Farha Bhuiyan, Taufika Islam Anee,
Masashi Inafuku, Hirosuke Oku and Masayuki Fujita

Additional information is available at the end of the chapter

http://dx.doi.org/10.5772/intechopen.68213

Abstract

Salicylic acid (SA) is an endogenous growth regulator of phenolic nature and also a signaling molecule, which participates in the regulation of physiological processes in plants such as growth, photosynthesis, and other metabolic processes. Several studies support a major role of SA in modulating the plant response to various abiotic stresses. It is a well-founded fact that SA potentially generates a wide array of metabolic responses in plants and also affects plant-water relations. This molecule also found to be very active in mitigating oxidative stress under adverse environmental conditions. Since abiotic stress remained the greatest constraints for crop production worldwide, finding effective approaches is an important task for plant biologists. Hence, understanding the physiological role of SA would help in developing abiotic stress tolerance in plants. In this chapter, we will shed light on the recent progress on the regulatory role of SA in mitigating abiotic stress.

Keywords: abiotic stress, antioxidant defense, oxidative stress, phytohormones, ROS signaling

1. Introduction

Abiotic stresses are a potential threat to agricultural productivity all over the world. Anthropogenic activities provoked the degradation of the agricultural system. Drought, excess soil salinity, extreme high, and low temperatures, metals/metalloids, ozone, UV-B radiation, nutrient (deficiency and excess) are the abiotic stresses which have increased many more times than previous due to anthropogenic activities [1, 2]. It has been projected that abiotic stresses may adversely affect 70% yield of staple food crops and decrease overall crop production by more than 50% [3, 4]. Thus, to improve plant performance and to reduce the loss of productivity caused by abiotic stress is vital. This can be implemented through various approaches and one of those is the application of exogenous phytoprotectant molecule.

Salicylic acid or orthohydroxy benzoic acid is ubiquitously distributed plant growth regulator [5]. Salicylic acid has positive effects on plant growth and developmental processes [5–7]. Research findings demonstrated its roles in seed germination, glycolysis, flowering, fruit yield [8], ion uptake and transport [9], photosynthetic rate, stomatal conductance (g_s), and in transpiration [10]. Salicylic acid can modulate antioxidant defense system thereby decreasing oxidative stress [11]. Photosynthesis, nitrogen metabolism, proline (Pro) metabolism, production of glycinebetaine (GB), and plant-water relations in abiotic stress affected plants were regulated by SA [12–14]. Induction of defense-related genes and stress resistance in biotic stressed plants have also been reported [15]. Moreover, exogenously applied SA showed putative positive effects on stressed plants [16–20]. Salicylic acid induced genes encoding chaperone, heat shock proteins (HSPs), antioxidants, and secondary metabolites of different types. Moreover, SA was involved in mitogen-activated protein kinase (MAPK) regulation, and in the expression [21]. There is no doubt about the vital roles of SA under abiotic stress condition. So, we will review and cover the area regarding the biosynthesis, involvement, and role of salicylic acid on abiotic stress affected plants.

2. Biosynthesis and metabolism of SA

Salicylic acid biosynthesis can occur through two distinct pathways viz. isochorismate (IC) pathway and phenylalanine ammonia-lyase (PAL) pathway (**Figure 1**). Both IC and PAL pathways are started with chorismic acid. Chorismic acid is the end product derived from the shikimic acid pathway in plastid [22–24]. Chorismic acid is converted to IC by isochorismate synthase (ICS) as reported in several plant species [25–27]. Isochorismate pyruvate lyase (IPL) supposed to catalyze the conversion of IC to SA (but the mechanism is not clear) [28].

In PAL pathway, deamination of phenylalanine is accomplished by the activity of PAL which generates *trans*-cinnamic acid. *Trans*-cinnamic acid is converted to intermediate product *ortho*-coumaric acid or producing the benzoic acid which later on produces SA [29–31].

After biosynthesis, SA can be modified into different other forms. Glucosylation of SA generates salicyloyl glucose ester (SGE) and salicylic acid 2-O-β-glucoside (SAG) where the activity of UDP-glucosyltrasferase is involved [32]. The SAG can be stored in the vacuole. Methylation of SA is occurred by SAM-dependent carboxyl methyltransferase to produce methyl salicylate (MeSA) [33]. After production, MeSA is transported to different parts of the plant. Through amino acid conjugation with SA, salicyloyl-L-aspartic acid (SA-Asp) generates (GH3-like phytohormone amino acid synthetase is proposed enzyme catalyze this reaction). SA-Asp can undergo through further catabolism [34]. The conversion of SA to SA-2-sulfonate is proposed to catalyze by sulfotransferase and this process is termed as sulfonation [35]. Hydroxylation of SA is responsible for the production of 2,5-dihydroxybenzoate (Gentisic acid) but the enzyme is unknown [36, 37].

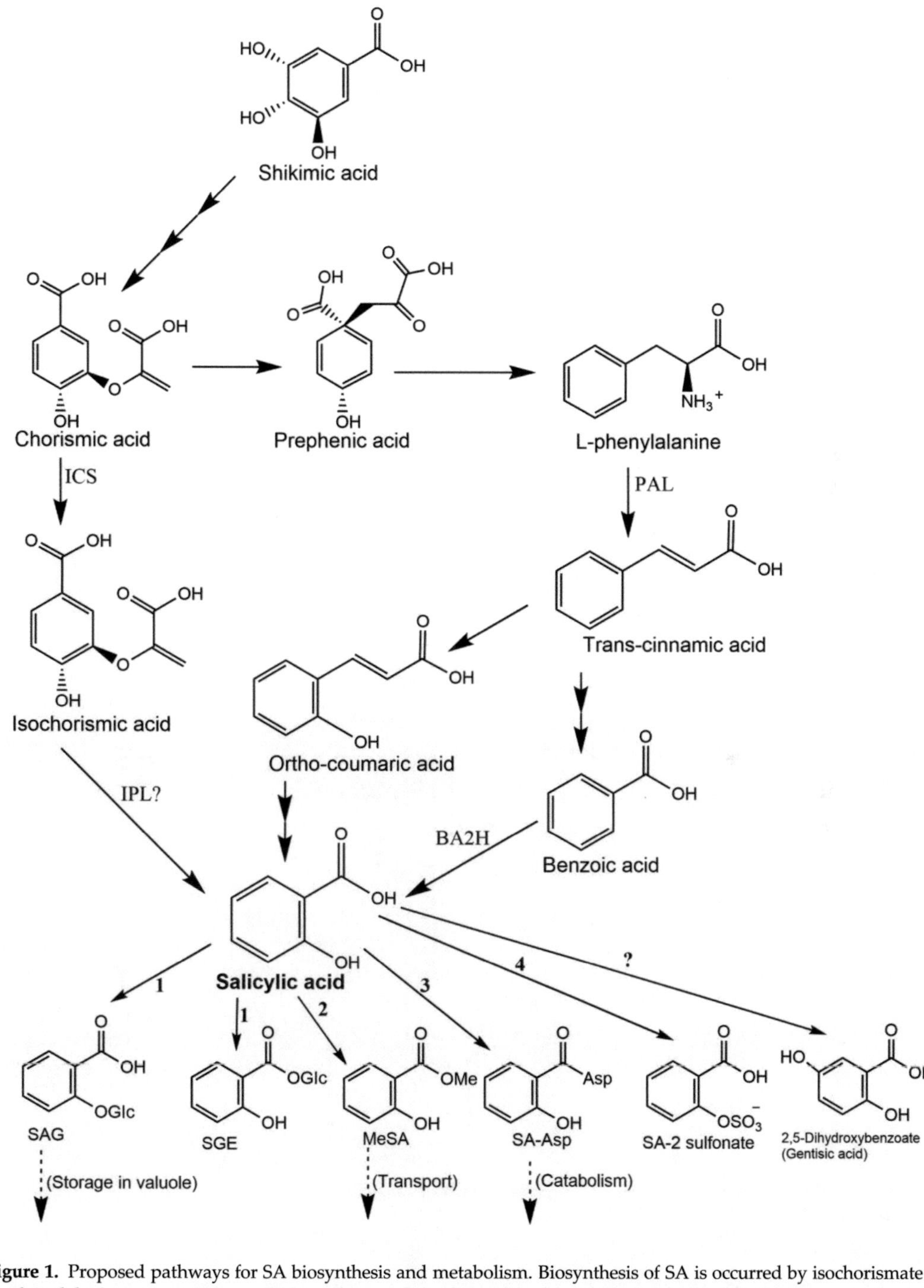

Figure 1. Proposed pathways for SA biosynthesis and metabolism. Biosynthesis of SA is occurred by isochorismate (IC) or phenylalanine ammonia-lyase (PAL) pathways. Salicylic acid is also metabolized into different forms. Isochorismate synthase (ICS), BA2H (benzoicacid-2-hydroxylase), IPL (isochorismate pyruvate-lyase); MeSA (methylsalicylate), SA-Asp (salicyloyl-L-aspartic acid), SAG (salicylic acid 2-O-β-glucoside), SGE (salicyloyl glucose ester) are involved in either or biosynthesis or metabolic pathway of SA. Here, 1 indicates UDP-glucosyltransferase, 2 indicates SAM-dependent carboxyl methyltransferase, 3 indicates GH3-like phytohormone amino acid synthetase, and 4 indicates sulfotransferase.

Modifications of SA often render it inactive but these modifications are also related to accumulation, function, and/or mobility. Glucosylation inactivates SA and allows vacuolar storage. Methylation inactivates SA and increases its membrane permeability, volatility which is vital for long-distance transport of this defense signal. Amino acid conjugation of SA is involved in SA catabolism [37].

3. Salicylic acid and abiotic stress tolerance

As a phytohormone, the role of SA in regulating plant growth and development is well known. The role of SA in mitigating abiotic stress has widely been studied since last few decades (**Tables 1–4**). A large volume of research reports indicate that both endogenous SA synthesis and exogenous application enhance plants tolerance to salinity [38–42], drought [43–45], extreme temperature [46–49], toxic metal and metalloids [50–53], and others [54–58]. Exogenous SA showed enhanced plant growth, photosynthesis, and decreased ROS production under various abiotic stresses (**Tables 1–4** and **Figure 2**).

3.1. Salinity

Among the prevailing catastrophic abiotic stresses, salinity or salt stress can be considered as the most devastating one. It shows enormous negative effects, both direct and indirect, on morphological, physiological and biochemical attributes of plants. When plants are exposed to

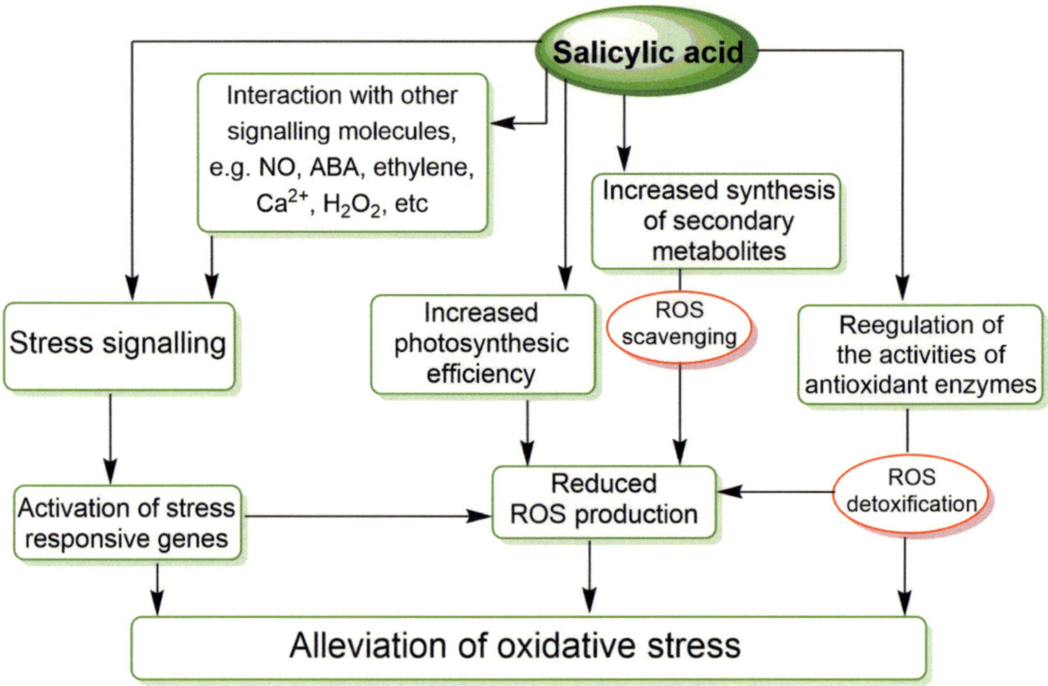

Figure 2. Some possible ways of SA-induced oxidative stress tolerance to plants.

Plant species	Salinity level	Effect of salinity	SA application	Protective effects	Reference
V. radiata L. cv. Pusa Vishal	100 mM NaCl, 20 d	• TBARS and H_2O_2 contents increased 2.5 and 4-times respectively, compared to control • Na^+ and Cl^- accumulation increased more than 3-times compared to control • Activity of GR, contents of GSH and GSSG increased compared to control • Net photosynthesis, g_s, C_i, carboxylation efficiency, WUE, and plant dry mass decreased compared to control • Content of methionine, GB, and ethylene evolution increased	Spraying 0.5 mM SA, 15 d	• TBARS and H_2O_2 contents decreased 1.5-times and 2.5-times respectively, compared to salt-treated plant • Reduced Cl^- and Na^+ contents by 50 and 39.8% compared to salt-treated plant • GR activity and GSH content increased but GSSG content decreased compared to stressed plant • P_n, g_s, C_i, carboxylation efficiency, WUE, and plant dry mass decreased compared to salt-treated plant • Content of methionine, GB and ethylene evolution decreased compared to stressed plant	Khan et al. [17]
T. aestivum cv. S-24 and MH-97	150 mM NaCl, 30 d	• Reduced shoot fresh and dry mass, and leaf area in both cultivars • Reduced grain yield plant^{-1}, 100-grain weight and number of spikelets in both cultivars • Reduced the net CO_2 assimilation rate (A), transpiration rate (E), g_s, C_i and WUE in both cultivars	0.25, 0.50, 0.75 and 1 mM SA in growth media, 30 d	• Increased fresh and dry masses of both shoot and root at 0.25 mM SA under saline condition • 0.25 mM SA reduced salt-induced damage in grain yield, 100-grain weight and number of grains of S-24 but, in case of MH-97 grain yield slightly improved with 0.50 mM SA application • SA increased A in S-24 at 0.25 mM concentration but in MH-97 at higher concentrations (0.75 and 1 mM)	Arfan et al. [59]
L. esculenta cv. DPL-62	100 mM NaCl, 10 d	• Decreased germination percentage, shoot length, root length, FW and DW • Free Pro and GB content increased in both shoot and root • Activities of P-5-CR and γ-glutamyl kinase increased but Pro oxidase decreased	0.5 mM SA in growth media, 10 d	• Improved germination percentage, shoot length, root length, FW and DW compared to salt stresses plants • Increased rate of free Pr and GB content was higher in shoot than root under salt stress • Activities of P-5-CR and γ-glutamyl kinase further increased and Pro oxidase decreased	Misra and Saxena [61]

Plant species	Salinity level	Effect of salinity	SA application	Protective effects	Reference
G. jamesonii L. cv. Amaretto	100 mM NaCl, 15 d	• EL and MDA content increased • Activities of SOD, POD, CAT and APX increased • Higher Pro content	0.5 mM SA pretreatment, 2 d	• EL and MDA content decreased • Activities of SOD, POD, CAT and APX further increased • Lower Pro content	Kumara et al. [39]
B. juncea L. cv Pusa Jai Kisan	100 mM NaCl, 30 d	• TBARS and H_2O_2 contents increased 2.5-times and 3.8-times respectively, compared to control • Increased Na^+ and Cl^- contents in leaves • Increased activities of DHAR, APX and GR by 30, 217 and 79%, respectively compared to control • Activities of ATPS and Serine acetyl transferase (SAT) and cystein (Cys) contents increased by 30, 23, and 70%, respectively, but S content decreased by 32% compared to control • Increased DHA, GSH, and GSSG contents but, decreased AsA content • Reduced net photosynthesis, g_s and C_i by 40.0, 26.4, and 41.3%, respectively, compared to control	0.5 mM SA spray, 15 d	• TBARS and H_2O_2 contents were decreased significantly • Reduced Na^+ content by 36% and Cl^- content by half compared to salt treated plants • Activities of DHAR, APX and GR further increased by 54, 248 and 111%, respectively compared to control • Activities of ATPS and SAT, and contents of Cys and S increased by 87, 76, 128, and 63% respectively, compared to control • GSH content further increased while DHA and GSSG both were reduced, AsA content increased • Limited the decreases in the above characteristics to 22, 19 and 25% respectively, compared to control	Nazar et al. [41]
S. lycopersicum Mill. L. cv. Rio Fuego	100 mM NaCl	• Higher accumulation of ABA in both leaves and root • Ethylene production increased • Reduced net CO_2 fixation rate	Pretreated with 0.01 μM and 100 μM SA, 21 d	• Lower accumulation of ABA compared to stressed plants • Prevented higher production of ethylene • Net CO_2 fixation rate enhanced	Horváth et al. [62]
Hordeum vulgare L. cv. Gustoe	150 mM NaCl, 14 d	• Shoot FW and height decreased by 30 and 36% respectively, compared to control • Photosynthetic pigment contents decreased by 57% compared to control • MDA content increased by 40% compared to control	50 μM SA, 14 d	• Shoot FW and height increased compared to NaCl treated plants • Photosynthetic pigment contents decreased by only 39% compared to control • MDA content was lower compared to salt-stressed plants • Na^+ content decreased and K^+ content increased	Fayez and Bazaid [63]

Plant species	Salinity level	Effect of salinity	SA application	Protective effects	Reference
		• Na^+ content increased and K^+ content decreased compared to control		compared to plants exposed to salt stress	
V. radiata L. cvs. Pusa Vishal and T44	50 mM NaCl, 30 d	• Increased root and shoot Na^+ and Cl^- contents • Decreased NR activity and N content • Leaf ATPS activity, S, H_2O_2, MDA contents and EL increased • Decreased water potential and osmotic potential by 69 and 52% in Pusa Vishal and 70 and 69% in T44, respectively compared to control • SOD, APX, GR activities and GSH, GSSG contents increased • g_s, C_i and chl fluorescence decreased	0.5 mM SA spray, 15 d	• Reduced root and shoot Na^+ and Cl^- contents • Increased NR activity and N content • Leaf ATPS activity and S content further increased, but H_2O_2, MDA contents, and EL decreased • Enhanced the water and osmotic potential by 54 and 34% in Pusa Vishal and 33 and 15% in T44 in comparison to control • SOD activity and GSSG content decreased, but APX and GR activities and GSH content further increased • g_s, C_i, and chl fluorescence increased	Nazar et al. [12]
Torreya grandis	0.4% NaCl, 60 d	• Reduced the dry mass of shoots and roots by 29 and 25%, respectively • RWC decreased by 16.70% but Pro content increased • P_n, transpiration rate (T_r) and g_s increased but C_i decreased • Chlorophyll content decreased by 24% • Activities of SOD, CAT, and POD decreased markedly • EL and MDA content increased	0.5 mM SA, 30 d	• Increased the dry mass of the shoots and roots by 16.8 and 18.2%, respectively • Both RWC and Pro content increased compared to stressed plants • P_n, T_r, and g_s decreased and C_i increased • Reduced the decrease of chl content by 17% • Increased the activities of SOD, CAT, and POD by 13, 38, and 19%, respectively • Reduced the EL rate and MDA content	Li et al. [64]
S. lycopersicum cv. Rio Fuego	100 mM NaCl, 10 d	• Water potential decreased and root length increased • Decreased Chl content and increased MDA content • ABA accumulation decreased in shoot but increased in root	Pretreated with 0.1 and 10 μM SA, 21 d	• Water potential increased and root length decreased • Improved Chl content and reduced MDA content than stressed one • ABA accumulation increased in shoot but decreased in root	Szepesi et al. [38]
A. thaliana	100 mM NaCl, 14 d	• Reduced FW, DW and water content • Increased shoot Na^+ and reduced K^+ contents	50 μM SA pretreatment, 1 h	• Improved FW, DW and water content • Decreased shoot Na^+ and enhanced K^+ contents	Jayakannan et al. [65]

Plant species	Salinity level	Effect of salinity	SA application	Protective effects	Reference
Medicago sativa cv. Aragon	200 mM, 12 d	• Reduced the F_V/F_M ratio by 15% • Decreased the nodule mass about 60% • Activities of CAT, APX and GR reduced but SOD, POX, and DHAR increased	Pretreated with 0.1 and 0.5 mM, 2 d	• Reduced the F_V/F_M ratio by only 2% • Reduced the decreasing of the nodule mass • Only CAT activity further reduced, but the activities of APX, GR, SOD, POX and DHAR further increased	Palma et al. [20]
T. aestivum L. cv. Yumai 34	250 mM NaCl, 3 d	• GSH content increased but AsA content decreased	0.5 mM SA, 3 d	• GSH content further increased along with AsA content	Li et al. [66]
Z. mays L., Hamidiye F1	40 mM NaCl, 56 d	• Membrane permeability and MDA content increased • Higher Na^+ and Cl^- concentrations • Decreased N and increased P concentrations	Soil incorporated with 0.5 mM SA, 56 d	• Ameliorated the deterioration of membrane and reduced MDA content • Lower accumulation of Na^+ And Cl^- ions • Increased N and decreased P concentrations	Gunes et al. [60]
S. lycopersicum cv. Rio Fuego	100 mM NaCl, 7 d	• Activities of APX, GR increased and SOD, CAT decreased • Both total ascorbate (AsA) and GSH contents increased in leaves and roots	Pretreated with 0.1 μM or 100 μM SA, 21 d	• Activities of APX, GR further increased along with SOD and CAT • Further increase of total AsA and GSH	Tari et al. [42]

Table 1. Salicylic acid–mediated tolerance of different plant species to salinity stress.

salt, they not only face osmotic and ionic stresses, but also water stress and other subsequent stresses may emerge. These ultimately reduce the quality and quantity of the desired yield. However, good news is there are certain species that show some tolerance mechanisms and also some protectants that can help plants to develop tolerance against the salt stress. In the recent era where global warming and rising of sea level are the most alarming issues, these can be promising facts to be considered for further research. There are a number of studies that prove the protective roles of SA against salt stress in many plant species (**Table 1**).

Salicylic acid has been proved to have effective roles on enhancing the germination percentage, shoot and root length, fresh weight (FW) and dry weight (DW) of both shoot and root of plants, uptake of beneficial ions, and also some antioxidant enzyme activities. It also has been proved to reduce the toxic ions uptake and accumulation in plants, membrane damage and transpiration rate, etc. Photosynthesis, growth, and yield were improved and oxidative damages were ameliorated with the application of effective concentrations of SA in different plant species. To demonstrate the role of SA in alleviating the salt stress-induced damage, an experiment was conducted by Arfan et al. [59] with two *Triticum aestivum* varieties, of which one is salt-tolerant (S-24) and another one is salt-sensitive (MH-97). They applied different levels of SA starting from 0.25 to 1.00 mM and created salt stress with 150 mM NaCl in the

Plant species	Drought condition	Effect of drought	SA application	Protective effects	Reference
T. aestivum. cvs. Giza164 and Gemaza 1	Drainage (control, 1/3 field capacity, 2/3 field capacity), 14 d	• Shoot and root DW and mineral content decreased • Increase of sugar, protein in shoot and root of plants	Seed soaking with 0.5 mM SA	• Higher accumulation of sugars, protein and mineral • Increased DW	El Tayeb and Ahmed [77]
B. juncea L. cv. BARI Sharisha 11	10 and 20% PEG, 48 h	• Increased MDA, H_2O_2 and Pro levels • Decreased leaf RWC, chl content • Decreased AsA content and increased in GSH and GSSG contents • GR, APX, GST activities increased • DHAR and Gly I activities decreased	Foliar spray with 50 μM SA	• Increased leaf RWC, chl, AsA and GSH contents • Decreased the GSSG content and maintained a higher ration of GSH/GSSG • Increased the activities of MDHAR, DHAR, GR, GPX, CAT, Gly I, and Gly II • Decreased H_2O_2 content and lipid peroxidation	Alam et al. [44]
Cymbopogn flexuosus Steud. Wats. cvs. Neema and Krishna	Water has been reduced to 75 and 50% of field capacity	• Reduced growth • Increased activities of nitrate reductase, carbonic anhydrase, and PEP carboxylase • Increased EL, Pro and free amino acid contents	Foliar application of 10 μM SA	• Improved growth parameters • Modulated the activities of nitrate reductase, carbonic anhydrase, and PEP carboxylase • Decreased EL, Pro and free amino acid contents	Idrees et al. [76]
T. aestivum	PEG 15%, 3 d	Decreased growth	0.5 mM SA, 3 d	• Overexpressed proteins associated with signal transduction, stress defense, photosynthesis, carbohydrate metabolism, protein metabolism, and energy production • Improved growth and drought tolerance in wheat seedlings	Kang et al. [43]
Satureja hortensis	1/3 and 2/3 of field capacity.	• Decreased sugar and protein contents • Increased Pro and lipid peroxidation	1.0 and 3.0 mM SA	• Increased sugar, protein and Pro contents • Decreased lipid peroxidation	Yazdanpanah et al. [78]
C. setosa	Kept in PEG, 4 h	• Increased leaf rolling	SA pretreatment, 14 h	• Increased Pro, sugar, activities of SOD, APX, CAT, GPX and NOX (NADPH oxidase)	Demiralay et al. [45]

Plant species	Drought condition	Effect of drought	SA application	Protective effects	Reference
		• Decreased antioxidant enzyme activities		• SA conferred drought tolerance mediated by H_2O_2 • Reduced leaf rolling	
Z. mays	Withholding water, 5 d	• Decreased RWC, leaf MSI, chl and K content	100, 150, 200 ppm	• Increased RWC, leaf MSI, chl, and K contents	Rao et al. [79]
Musa acuminata cv. 'Berangan', AAA	1, 2 and 3% of PEG in vitro, 60 d	• Slight increase in proliferation rate, FW • Decreased RWC and chl content	1.0, 2.0, and 3.0 mM SA	• Increased Pro content • Reduction in H_2O_2 and MDA contents	Bidabadi et al. [80]
H. vulgare L. *cv* Nosrat	40% FC	• Drought stress decreased the dry mass and net CO_2 assimilation rate	Spraying with 500 μM SA, 15 d	• Increased the dry mass, net CO_2 assimilation rate and g_s	Habibi [81]
O. sativa L. cv. Super-Basmati	Water was reduced to 50% of field capacity	• Increased H_2O_2, MDA, and relative membrane permeability	100 mg L^{-1} SA	• Increased tissue water potential, increased synthesis of metabolites and enhanced capacity of the antioxidant system	Farooq et al. [82]
T. aestivum L.	20 % PEG, 24 h	• Decreased MSI, increased total soluble sugar and soluble protein content but decreased yield	10 μM SA, 24 h	• Increased MSI, increased total soluble sugar and soluble protein content • Increased yield	Khan et al. [73]
T. aestivum L. cv. Hassawi	Reduced water to 60 and 30% field capacity	• Decreased content of photosynthetic pigments • Increased soluble carbohydrate, protein, and Pro • Decreased insoluble carbohydrates and proteins • Free amino acids were significantly increased in roots, while it was decreased in shoots	50 ppm SA	• Stimulated growth, photosynthetic pigments and accumulation of soluble and insoluble carbohydrates and proteins	Azooz and Youssef [83]

Table 2. Salicylic acid–mediated tolerance of different plant species to drought stress.

growing media. Both SA and NaCl were exposed to plants from the very beginning and data were taken at 30 days after sowing (DAS). Their results showed that at lower concentration

Plant species	Temperature and duration	Damaging effects	SA dose and duration	Protective effects	References
T. aestivum	HT (30°C, 2 h)	• Decreased FW • Increased total soluble protein and total RNA contents • Decreased soluble starch synthase activity	100 mM, foliar spraying, 3 times at 4 h interval	• Improved FW • No significant differences in the total soluble protein content • Further increased total RNA content • Increased soluble starch synthase activity	Kumar et al. [46]
Z. mays	HT (40 ± 1°C, 2 h)	• Reduced dry biomass • Increased MDA, H_2O_2 but decreased Pro contents • APX and GR activities increased but CAT and SOD activities decreased	10–800 μM, foliar spraying, 2 h	• Improved dry biomass content • Reduced MDA, H_2O_2 but increased Pro contents • Further improved CAT, SOD and POX activities	Khanna et al. [48]
L. lycopersicum	HT (32/26°C, 12/12 h, day/night)	• Severely reduced germination and plant growth • Negatively affected reproductive parameters • Decreased TSS (total soluble solid), vitamin and lycopene contents	0.25, 0.5, and 0.75 mM, seed priming	• Reduced germination time and increased germination percentage • Improved reproductive parameters. Increased TSS, vitamin and lycopene contents	Singh and Singh [49]
T. aestivum	HT (40°C, 6 h)	• Increased Pro and γ-glutamyl kinase (GK) but reduced Pro oxidase contents • Increased TBARS and H_2O_2 contents • Negatively affected the net photosynthesis, Rubisco activity, chl and WUE of the plant	0.5 mM, foliar spraying	• Further increased Pro and GK and reduced Pro oxidase contents • Reduced TBARS and H_2O_2 contents • Counteracted the negative effects on net photosynthesis, Rubisco activity, chl and WUE of the plant	Khan et al. [1]
Digitalis trojana Ivanina	HT (45°C, 2 or 4 h)	• Lowered CAT and SOD activities • Increased accumulation of Pro • Increased total phenolic and flavonoid contents	150 μM, pretreatment, 4 h	• Increased CAT and SOD activities • Further increased Pro content • Further increased phenolic and flavonoid contents	Cingoz and Gurel [47]
Vitis vinifera	HT (43°C, 5 h)	• Sharply declined P_n and g_s • Lowered Rubisco activity	100 μM SA, pretreatment, 24 h	• Alleviated the harmful effect on P_n • Counteracted the negative effect on Rubisco activation state	Wang et al. [95]

Plant species	Temperature and duration	Damaging effects	SA dose and duration	Protective effects	References
Matricaria chamomila	HT, Min (10.1–28.2°C), Max (21–44°C), 8 months	• Reduced plant height, capitol diameter, fresh flower weight, dried flower weight etc. • Decreased total chl content • Essential oil content was not significantly affected	1, 10, 25 and 100 mg L^{-1}, foliar spraying	• Improved plant height, capitol diameter, fresh flower weight, dried flower weight etc. • Increased total Chl content • Improved essential oil content	Ghasemi et al. [96]
Cucumis sativa	HT (40°C, 36 h)	• Highly increased EL • Increased H_2O_2 and TBARS contents • Improved SOD, CAT, DHAR, GPX, APX and GR activities	1 mM SA, foliar spraying, 12 h	• Decreased EL • Reduced H_2O_2 and TBARS contents • Improved SOD, DHAR, GPX, APX and GR activities but inhibited CAT activity	Shi et al. [97]
B. juncea	HT (30 or 40°C, 24 h)	• Decreased root length, shoot length, FW and DM of the plant • Decreased P_n, g_s, C_i, WUE and SPAD value • Increased activities of CAT, POD SOD, and Pro accumulation • Decreased N, P, K contents in leaves	10 μM, foliar spraying	• Root length, shoot length, DW and FW of plant increased • Increased P_n, g_s, C_i, WUE and SPAD value • Further enhancement of CAT, POD SOD, and Pro accumulation • Improved N P, K contents in leaves	Hayat et al. [98]
Musa acuminata	Chilling (5°C, 3 d)	• Reduced SOD, CAT, APX activities but improved POX activity • Increased accumulation of H_2O_2	0.5 mM, pretreatment, 1 d	• Increased SOD, CAT and APX activities but unaffected POX activity • Decreased overproduced H_2O_2	Kang et al. [99]
T. aestivum	Chilling (3 °C, 48–72 h)	• Reduced chl, CO_2 assimilation and rate of respiration • RuBisCO activity decreased • Decreased SOD • Increased glycolate oxidase (GO) and CAT activities • Highest MDA content found	500 μM, foliar spraying, 24 h	• Improved Chl content and rubisco activity • Enhanced CAT, APX, POX, and glycolate oxidase (GO) activities but GR activity found not affected • Reduced MDA content	Yordanova and Popova [100]
V. vinifera	HT (38 ± 0.5°C, 12 h)	• Reduced H^+ and Ca^{2+}-ATPase activities • The appearance of cerium phosphate grain found	100 μM, pretreatment, 6 h	• Increased H^+ and Ca^{2+}-ATPase activities • Remained cerium phosphate grain	Liu et al. [101]

Plant species	Temperature and duration	Damaging effects	SA dose and duration	Protective effects	References
V. vinifera	Freezing (−3°C, 3 h) and HT (44°C, 3 h)	• Increased EL and TBARS content • Decreased activities of GR, APX, MDHAR and DHAR • Increased AsA/GSH ratio • Reduced cytosolic Ca^{2+} homeostasis	2 μM, pretreatment, 1, 3, 6, or 12 h	• Reduced EL and TBARS content • Increased activities of GR, APX, MDHAR but reduced DHAR and ratio of AsA/GSH pool • Increased cytosolic Ca^{2+} homeostasis	Wang and Li [102]
Poa pratensis	HT (46°C, 72 h)	• Overproduction of $O_2^{\bullet-}$ and H_2O_2 in a time-dependent manner • Enhanced CAT and SOD activities in a time-dependent manner	0.1, 0.25, 0.5, 1.0, and 1.5 mM, foliar spraying	• Reduced $O_2^{\bullet-}$ • H_2O_2 generation in a time-dependent manner • Significantly increased CAT and SOD activities in a time-dependent manner	He et al. [103]
O. sativa	HT (27–32°C, night ambient temp)	• Spikelet fertility and grain size reduced • Damaged membrane and resulted oxidative damage	1 mM, foliar spraying	• Improved grain characteristics and spikelet fertility • Counteracted oxidative damage • Enhanced antioxidant enzymes activities	Mohammed and Tarpley [104]
Prunus persica	Chilling (0°C, 28 d)	• Increased TBARS content, chilling injury index and decay index • Lowered APX and GR activities and heat shock protein 101 (HSP101) expression	0.35, 0.7, and 1 mM, 5 min	• Maintained optimum level of TBARS content, chilling injury index and decay index • Enhanced reduced to oxidized AsA and GSH ratios • Higher APX and GR activities and heat shock protein 101 (HSP101) expressions	Wang et al. [105]
Rhododendron	HT (38/30°C (day/night, 6 d)	• Resulted in brown, withered and defoliated leaves • Decreased chl and total soluble protein contents • Decreased activity of POX and SOD • MDA and H_2O_2 contents increased	0.5, 1.0, and 2.0 mM spraying of leaves, 3 d	• Lowered the leaf damage rate • Highly declined Chl and total soluble protein contents • Increased POX and SOD activities • Reduced MDA and H_2O_2 contents	Shen et al. [106]
V. radiata	HT (50°C, 3 h)	• Increased lipid peroxidation, H_2O_2, and EL • Reduced CAT, APX, POD and GSH contents	0.5 and 1 mM, foliar spraying	• Decreased lipid peroxidation, H_2O_2, and EL • Increased CAT, APX, POD and GSH contents	Saleh et al. [107]

Plant species	Temperature and duration	Damaging effects	SA dose and duration	Protective effects	References
		• Increased endogenous SOD and SA activities		• Enhanced defense system and radical scavenging mechanism	
Z. mays	Chilling (15°C)	• Disturbed seedling emergence, root and shoot growth, FW, and DW • Lowered RWC and increased membrane permeability	50, 100, and 150 mg L^{-1}, seed priming, 24 h	• Improved seedling emergence, root and shoot growth, FW and DW • Increased RWC and decreased membrane permeability • Highly activated CAT, SOD and APX activities	Farooq et al. [108]
Brassica sp.	HT (40–55°C, 3 h)	• Hampered seedling growth • Increased EL • Increased CAT, POX enzyme activities	10 and 20 μM, pretreatment, 2 h	• Conferred protection of membrane • Reduced EL • Increased FW, DW, and total soluble sugar • Increased CAT, POX, and appearance of heat shock proteins	Kaur et al. [3]

Table 3. Summary of the protective roles of exogenous SA in mitigating extreme temperature-induced damages in different crop plants.

(0.25 mM) SA could reduce the negative effects of salt on the shoot and root FW and DW in tolerant variety, but this was true at higher concentration (0.75 mM) for the sensitive one. However, yield attributes such as grain yield and 100-grain weight were increased in both the varieties at lower concentrations (0.25 and 0.50 mM) SA under salt stress. Similarly, application of SA also increased the water use efficiency (WUE) of those varieties. Similarly, Nazar et al. [12] also chose two such varieties of *Vigna radiata* cvs. Pusa Vishal (salt-tolerant) and T44 (salt-sensitive). They also used different concentrations of SA (0.1, 0.5, and 1.0 mM) against 50 mM NaCl stress and 0.5 mM was concluded as the most suitable concentration for both varieties irrespective of their tolerance ability. At this concentration, *V. radiata* seedlings could reduce the accumulation of toxic Na^+ and Cl^- ions, and increase S and N uptake and nitrate reductase activity. Salicylic acid application also enhanced the water and osmotic potential which was higher in the tolerant one. Stomatal conductance (g_s), intercellular CO_2 concentration (C_i), and chlorophyll (chl) fluorescence were increased along with leaf ATP-sulfurylase activity. However, mainly the reduction of electrolyte leakage (EL), malondialdehyde (MDA), H_2O_2, oxidized glutathione (GSSG) contents, superoxide dismutase (SOD) activity and enhancement of reduced glutathione (GSH) content, and ascorbate peroxidase (APX) and glutathione reductase (GR) activities prove the role of SA in reducing salt-stress damages. Similar results were observed with *V. radiata* seedlings. These seedlings were exposed to 100 mM NaCl at 10 DAS and then was spraying SA (0.5 mM) at 15 DAS. At 30 DAS some parameters were monitored related to gas exchange e.g., net photosynthesis (Pn), gs, and Ci; and also carboxylation efficiency, WUE, and plant dry mass [17]. Application of SA increased P_n, g_s, and C_i by 17.9, 19.2, and 23.5%, respectively, under salt stress condition. It also enhanced carboxylation

Plant species	Toxic metals/ metalloids	Doses and duration	Toxic effects	SA doses and duration	Protective effects	References
O. sativa	As	25 and 50 μM Na_2HAsO_4, 7 d	• Decreased growth and biomass production • Enhanced MDA and H_2O_2 contents	100 μM, pretreatment, 7d	• Reverted the growth inhibition • Reduced oxidative stress by reducing MDA and H_2O_2 contents	Singh et al. [50]
A. thaliana	As	100 μM Na_2HAsO_4, 14 d	• Decreased plant biomass and chl contents • Increased lipid peroxidation and antioxidant enzymes (APX, CAT, SOD) activities	250 μM, pretreatment, 14 d	• Alleviated the toxic effects of As by restoring growth and chl content • Decreased lipid peroxidation and downregulated activities of APX, CAT, and SOD	Odjegba [115]
T. aestivum	Cd	0.01, 0.1 and 1 mM Cd-$(OOH)_2$, 24 h	• Increased MDA and EL percentage • Reduced FW and DW • Enhanced accumulation of ABA and dehydrins	50 μM, 24 h, pretreatment	• Declined MDA content and EL • Recovered growth • Activated phenylalanine ammonia-lyase (PAL) enzyme activity • Further increased low molecular weight dehydrins by 1.5 folds	Shakirova et al. [53]
H. vulgare	Cd	15 μM $CdCl_2$, 30 min	• Reduced root growth markedly • Stimulated activity of LOX, GPX and enhanced accumulation of IAA • Increased generation of ROS in the root apex	0.25 or 0.5 mM SA, pretreatment, 10 min	• Alleviated Cd-induced root growth inhibition • Reduced IAA-induced LOX and GPX activity • Inhibited ROS generation in roots	Tamás et al. [51]
Lolium perenne	Cd	100 μM $CdCl_2$, 14 d	• SOD, APX, and CAT activity decreased drastically in both shoots and roots. Increased accumulation of $O_2^{\bullet-}$, H_2O_2 and increased MDA content	100, 200, 300, and 400 μM, 14 d	• Increased antioxidant enzyme activities and chl content • Increased mineral uptake but decreased Cd uptake • Decreased accumulation of MDA and H_2O_2	Bai et al. [116]
Phaseolus vulgaris	Cd	0.25, or 0.50 mM $CdCl_2$, 2–3 d	• Increased Pro and Cd^{2+} ion accumulation • Enhanced EL and lipid peroxidation	1 mM, foliar spraying,50, 36, and 22 d	• Decreased EL and lipid peroxidation • Reduced activities of antioxidant enzymes, such as SOD, CAT, APX, and GR	Wael et al. [117]

Plant species	Toxic metals/ metalloids	Doses and duration	Toxic effects	SA doses and duration	Protective effects	References
			• Increased activities of SOD, CAT, APX, and GR			
T. aestivum	Cd	100, 400 and 1000 μM $CdCl_2 . 2.5 H_2O$, 30 d	• Inhibited root growth and enhanced Cd accumulation in roots • Decreased RWC, chl content, and CO_2 fixation • Increased MDA, H_2O_2 and Pro contents • Altered root and chloroplast ultrastructure	500 μM, pretreatment, 20 h	• Reversed root growth inhibition • Ameliorated the adverse effects on RWC, chl content, and CO_2 fixation • Reduced MDA, H_2O_2 and Pro contents • Recovered chloroplast and root ultrastructures	Moussa and El-Gamal [118]
Linum usitatissimum	Cd	50 and100 mM $CdCl_2$, 4 d	• Inhibited growth and nutrient absorption • Enhanced MDA content and EL • Reduced total lipid and chl contents	250 and1000 μM, presoaking of grains, 8 h	• Alleviated growth inhibition and nutrient absorption • Ameliorated the enhanced MDA content and EL • Alleviated the harmful effect on total lipid and Chl contents	Belkhadi et al. [119]
Phaseolus aureus	Cd	50 and 100 μM $CdCl_2$, 3 or 6 d	• Enhanced H_2O_2 and $O_2^{\bullet -}$ production in the root • Increased TBARS content and relative EL rate • Increased antioxidant enzymes such as SOD, CAT and APX activities	100 μM SA, seed soaking, 16 h	• Significantly decreased H_2O_2, $O_2^{\bullet -}$ production in the root • Enhanced TBARS content and relative EL • Further increased • SOD, CAT and APX activities	Zhang et al. [120]
T. aestivum	Cd	500 and 1000 μM $CdCl_2$, 3 d	• Reduced growth, chl content, RWC and SOD, CAT, POX activities • Enhanced Pro, EL and Cd contents	500 μM, seed soaking, 12 h	• Mitigated adverse effects of Cd on chl, RWC and SOD, CAT and POX activities • Alleviated damaging effects on EL • Improved leaf anatomy and reduced uptake of Cd	Agami and Mohamed [121]
Z. mays	Cr	50 mg L^{-1} $K_2Cr_2O_7$, 7 d	• Decreased growth, photosynthetic pigments, CHO metabolism	100 μM L^{-1}, foliar application, 15 d	• A significant decline in MDA, H_2O_2, Pro and Cr contents	Islam et al. [52]

Plant species	Toxic metals/ metalloids	Doses and duration	Toxic effects	SA doses and duration	Protective effects	References
			• Increased MDA, H_2O_2, Pro and Cr contents • Decreased CAT but increased SOD and POD activities		• Improved plant growth, photosynthetic pigments • Reduced oxidative stress by upregulating antioxidant enzymes (CAT, SOD, POD) activities	
Catharanth roseus	Ni	50, 100, and 150 mg kg^{-1} $NiSO_4 \cdot 6H_2O$	• Significantly reduced growth, photosynthetic pigments and activities of carbonic anhydrase and nitrate reductase. Increased SOD, CAT, POX activities, EL and Pro contents	10 μM, foliar spraying, 4 sprays at 10 d intervals from 30 DAS	• Restored plant growth processes • Further increase of antioxidant enzymes (SOD, CAT, POD) activities	Idrees et al. [122]
B. napus	Ni	0.5 mM $NiCl_2 \cdot 6H_2O$, 10 d	• Resulted leaf chlorosis and necrosis • Decreased growth and leaf chl content • LOX, MDA, H_2O_2, and Pro content increased • Reduced CAT, APX and GPX activity	0.2 mM SA, cotreatment, 10 d	• Decreased leaf chlorosis and necrosis • Improved growth and leaf chl content • A significant decline in LOX, MDA, H_2O_2 and Pro contents • Increased CAT, APX and GPX activities	Kazemi et al. [123]
O. sativa	Cu	75 or 150 μM $CuSO_4$, 48 h	• Highly increased MDA, $O_2^{\bullet -}$ H_2O_2 contents and LOX activity • Reduced Chl and leaf RWC • Reduced AsA GSH, nonprotein thiol and pro contents in roots	100 μM SA, pretreatment, 24 h	• Curtailed Cu-induced enhancement of MDA, $O_2^{\bullet -}$, H_2O_2 contents and LOX activity • Increased Chl, leaf RWC, AsA, GSH, nonprotein thiol contents • Enhanced antioxidant enzymes (SOD, CAT, APX, GR, GPX, DHAR and GST) activities	Mostofa and Fujita [124]
Zygophyllum fabago	Pb	0.75 mM Pb $(NO_3)_2$,7 d	• Downregulated CAT, APX, GR but upregulated DHAR activity • Increased MDA and EL	0.5 mM SA, pretreatment, 24 h	• Increased SOD activity • Reduced accumulation of Pb and chl contents	Lopez-Orenes et al. [125]
T. aestivum	Pb	10, 50, 100, 200 mg L^{-1} Pb^{2+}, 7 d	• Decreased chl content in a dose-dependent manner	200 mg L^{-1}, cotreatment, 7 d	• Alleviated the inhibitory effects on Chl and antioxidant	Song et al. [126]

Plant species	Toxic metals/ metalloids	Doses and duration	Toxic effects	SA doses and duration	Protective effects	References
			• Inhibited APX, CAT, MDA and Pro contents • POD, SOD, and soluble sugar contents were affected		enzymes activities. Lowered MDA and Pro contents • Reduced the adverse effects of Pb on POD, SOD, and soluble sugar contents	
P. pratensis	Cd	5,10, or 50 mM $CdCl_2$, 7 d	• Reduced growth, chl and nutrient (K, Ca, Fe, Mg) elements • Elevated MDA and H_2O_2 contents • Increased Cd uptake and accumulation	500 mM SA, pretreatment, 12 h	• Increased growth, chl, and nutrient (K, Ca, Fe, Mg) elements • Decreased MDA and H_2O_2 contents and Cd uptake • Marked increase in SOD, APX, and POD but decreased CAT activity	Guo et al. [127]
Spinacia oleracea	B	50 mg kg^{-1}, H_3BO_3	• Enhanced B Accumulation in plants • Decreased chl and anthocyanin contents • Increased MDA, H_2O_2 and stomatal resistance	0.5 mM kg^{-1}, cotreatment	• Decreased B accumulation • Improved chl and anthocyanin contents • Influenced antioxidant enzymes activity and stomatal resistance	Eraslan et al. [128]
Daucus carota	B	25 mg kg^{-1}, H_3BO_3	• Increased storage root diameter • Increased oxidative damage as indicated by increased MDA content • Lowered chl content	0.5 mM kg^{-1}, cotreatment	• Enhanced storage root DW • Increased anthocyanin and carotenoid contents • Controlled metal toxicity and pro accumulation in roots and shoots	Eraslan et al. [129]
B. juncea	Mn	3.0, 6.0, or 9.0 mM $MnCl_2$, 3 d	• Decreased growth, photosynthetic pigments, carbonic anhydrase activity, and water relations • Increased MDA, H_2O_2, Pro accumulation • Elevated antioxidant enzymes (CAT, SOD, POD) activity in a dose-dependent manner	10 μM, 14 d, foliar application	• Improved growth, photosynthetic pigments, carbonic anhydrase activity, and water relations • Lowered MDA, H_2O_2, and EL in a dose-dependent manner • Further accelerated activity of antioxidant enzymes (CAT, SOD, POD)	Parashar et al. [130]

Plant species	Toxic metals/ metalloids	Doses and duration	Toxic effects	SA doses and duration	Protective effects	References
C. sativus	Mn	600 μM $MnSO_4$, 11 d	• Caused stunted growth, severe chlorosis, a marked increase in Mn accumulation • Inhibited nutrients (Ca, Mg, Zn) absorption • Increased ROS production and lipid peroxidation	100 μM SA, co-treatment, 11 d	• Promoted growth and reduced Mn toxicity • Alleviated the effects on nutrients (Ca, Mg, Zn) absorption • Reduced Increased ROS production and lipid peroxidation • Increased SOD, POD, DHAR, GR, ASA and GSH activities but decreased CAT and POX activities	Shi and Zhu [131]
Glycine max	Al	30 μM $AlCl_3$, 12 h	• Inhibited root elongation • Accelerated SOD, POD, and APX activities in roots • Upregulated the CaM-like protein genes	10 μM, cotreatment, 12 h	• Restored root growth • Further enhanced SOD and POD activities • Upregulated CaM-like protein genes • Increased cytosolic Ca^{2+} • Concentration	Lan et al. [132]
Vallisneria natans	Pb	50 μM $Pb(NO_3)_2$, 4 d	• Decreased total chl content • Increased MDA, $O_2^{\bullet -}$ H_2O_2 contents • Increased CAT but no significant effect on APX and POD	10 or 100 μM, cotreatment, 4 d	• Decreased Chl and carotenoid contents • Inhibited the overproduction of MDA, $O_2^{\bullet -}$ and H_2O_2 • Decreased APX but increased CAT, DHAR and POD activities • Decreased NADPH oxidase, nonprotein thiols, AsA	Wang et al. [133]
Medicago sativa	Hg	10 μM $HgCl_2$, 24 h	• Increased Lipid peroxidation and TBARS accumulation • Slight increase in NADH oxidase, SOD and POD activities • Decreased APX and GR activities • Decreased AsA, GSH, and Pro accumulation	0.2 mM, pretreatment, 12 h	• Increased activities of NADH oxidase, APX, POD, GR but decreased SOD activity • Elevated AsA, GSH and Pro accumulation	Zhou et al. [134]
B. oleracea var. botrytis	Co, Ni, Cd, Cr, and Pb	0.25 M	• Retarded growth	50 and 100 mM, cotreatment	• Reversed all the toxic effects caused by HM except for Cr	Sinha et al. [135]

Plant species	Toxic metals/ metalloids	Doses and duration	Toxic effects	SA doses and duration	Protective effects	References
			• Increased MDA, nonprotein thiol, EL percentage and Pro contents • Increased POD and SOD activities		• In the case of Cr, SA accelerated the toxic effects and the plats died	
Pisum sativum	Cd	0.5, 1, 2, and 5 μM $CdCl_2$	• Decreased FW, CO_2 fixation, chl content and RuBPC activity • Increased MDA, Pro content, and EL percentage	500 μM, seed pretreatment, 6 h	• Restored FW, CO_2 fixation, chl content and RuBPC activity • Alleviated the effects on MDA, Pro contents, and EL percentage	Popova et al. [136]
Cannabis sativa	Cd	$CdCl_2$. 2.5 H_2O at 0, 25, 50, and 100 mg kg^{-1} sands	• Inhibited plant growth • Slightly reduced photosynthetic capacity • Increased Cd uptake	500 μM, seed soaking, 6 h	• Counteracted growth inhibition • Improved photosynthetic capacity • Reduced Cd uptake • Enhanced SOD and POD activities	Shi et al. [137]
H. vulgare	Zn^{2+}, Cu^{2+}, Mn^2, Cd^{2+}, Hg^{2+}, and Pb^{2+}, 7 d	0.1, 0.2, 0.5, and 1 mM Zn^{2+}, Cu^{2+}, Mn^2, Cd^{2+}, Hg^{2+}, and Pb^{2+}	• Inhibitory effects on SOD and CAT at higher dose	2 mM, cotreatment, 7 d	• Alleviated the harmful effects on antioxidant enzymes (CAT, SOD) activities	Song et al. [138]

Table 4. Summary of the protective roles of SA in mitigating toxic metal/metalloid–induced damages in different crop plants.

efficiency and WUE compared to salt-stressed plants. Plant dry mass was increased by 25.2% under salt condition compared to control plants. Meanwhile, *Zea mays* was tested with different levels of SA under salt stress, and positive roles of SA was demonstrated in ameliorating the membrane damage by reducing MDA content [60]. It can also decrease the accumulation of Na^+ and Cl^- ions and increase uptake of N and P and thus render tolerance to plants against salt stress. Another experiment was conducted with *Lens esculenta,* which included only four with *Lens esculenta,* which included only four treatments: nonsaline control (I), 0.5 mM SA (II), 100 mM NaCl (III), and the combination of 100 mM NaCl + 0.5 mM SA (IV). The results showed that growth parameters: germination (%), shoot and root length, FW and DW were improved in treatment (IV) compared to treatment (III). In addition, SA increased the free Pro and GB content in shoot and also the activities of pyrroline-5-carboxylate reductase (P-5-CR) and γ-glutamyl kinase which are the enzymes related to Pro anabolism. But, in contrast, it reduced the activity of Pro oxidase [61]. In the case of pretreatment with SA, it also showed some positive results. *Solanum lycopersicum* seeds pretreated with 10 μM SA improved the chl content and reduced MDA content under salt stress (100 mM NaCl) [38]. Higher accumulation of abscisic acid (ABA) in shoot and enhancement of water potential of SA-treated seedlings compared to the seedlings exposed to salt alone were also observed [38]. These results were supported by Horváth et al. [62] in the same plants with the equal concentrations of SA. Similarly, when pretreatment with SA (0.5 and 1.0 mM) was done, *Gerbera jamesonii* seedlings also showed positive results in salt stress (100 mM). Salicylic acid application reduced the EL, MDA and Pro contents and increased the activities of SOD, catalase (CAT), peroxidase (POD), and APX compared to salt stressed seedlings [39]. But, these effects were more acceptable in case of lower (0.5 mM) concentration of SA. Recently, Nazar et al. [41] again used SA (0.5 mM) to demonstrate the preventive role of it in *Brassica juncea* seedlings exposed to 100 mM of NaCl stress for 30 consecutive days. Application of SA reduced thiobarbituric acid reactive substances (TBARS) and H_2O_2 contents, also dehydroascorbate (DHA) and GSSG contents. It was found to increase the activities of dehydroascorbate reductase (DHAR), APX, and GR to a remarkable content. And most importantly, it reduced the toxic Na^+ and Cl^- uptake to almost half of the salt-stressed plants [41].

From the above-mentioned studies, the role of SA in alleviating salt stress can be considered as clear and concise. But, there are also some points to be considered as higher concentrations of SA may itself cause damage to plants [12, 59] and very lower concentrations may have a minimum effect [42, 62] against salt stress. So, the concentration of SA, application method and time, duration of salt stress and plant age are some of the important points to be considered while using SA against salt stress.

3.2. Drought

Drought stress is one of the most devastating abiotic stresses adversely affecting growth and developmental processes of the plant. Drought stress affects the physiological processes, brings biochemical changes, leads to the formation of secondary metabolites, significantly accumulates endogenous reactive oxygen species (ROS) and increases toxins (such as methylglyoxal). Drought stress hampering the reproductive development drastically reduces yield or productivity of plants [67].

Several studies demonstrated and proved the pivotal roles of SA in alleviating drought damage and improving drought stress tolerance in plants (**Table 2**). Salicylic acid pretreatment (0.5 mM) alleviated substantial water loss and its damaging effects on wheat seedlings that enhanced drought tolerance [43]. Pretreatment with SA upregulated 37 protein spots under drought stress which has been investigated through proteomics. Glutathione *S*-transferases, APX, and 2-cysteine peroxiredoxin were enhanced under drought stress. Enhancement of antioxidant defense system worked against the oxidative damage [43]. Proteins involved in ATP synthesis are also upregulated by SA under drought stress. Salicylic acid supplementation with drought also upregulated 21 protein spots, including RuBisCo and related enzymes [43]. In their other experiment, influential role of SA was also demonstrated on AsA-GSH cycle [68]. Exogenous SA supplementation enhanced the transcription of *GST1, GST2, GR*, and *MDHAR* genes during almost the entire drought period. The increase of *DHAR* was noticed at 12 h, *GPX1* at 48 h, phospholipid hydroperoxide glutathione peroxidase (*GPX2*) at 12 and 24 h, and glutathione synthetase (*GSHS*) at 12, 24, and 48 h of drought stress. Upregulation of transcription level of AsA-GSH cycle enzymes contributed to drought tolerance [68]. SA-accumulating (*siz1* and *cpr5*) genes were highly expressed in guard cells of drought which modulated movement of stomatal aperture in *Arabidopsis* plants. The generation of ROS was also modulated in this plant [69]. In tomato (*Lycopersicon esculentum*), SA treatment with drought has been demonstrated to protect the activity of nitrate reductase which helps to maintain the protein and nitrogen contents of the leaves, compared to the drought affected plant without SA addition. Photosynthetic parameters, membrane stability, water potential and activity of carbonic anhydrase were maintained by SA which also contributed to drought stress tolerance [70]. In sunflower, water stress-induced decrease in the yield and oil content. Salicylic acid (0.724 mM) application increased the Pro content, head diameter, number of achene, 1000-achene weight, achene yield, and oil yield of sunflower, compared to drought treatment alone [71]. The addition of acetyl SA in (0.1–1.0 mM) also improved drought tolerance of muskmelon seedlings [72]. Two wheat varieties viz. Wafaq-2001 and Punjab-96 were subjected to drought stress. Drought stress significantly decreased membrane stability index (MSI) and yield. Salicylic acid supplementation caused 37% increase in soluble sugars in Wafaq-2001 cultivar which was higher, compared to Punjab-96 cultivar. Salicylic acid also increased protein content and MSI in both cultivars with a higher increase in Wafaq-2001. The overall drought tolerance was higher in Wafaq-2001 after SA application which is evident from higher yield [73]. Exogenous addition of SA increased the activity of antioxidant enzymes which helped to alleviate the drought stress damage in *Ctenanthe setosa* [74]. Mustard (*B. juncea* L. cv. BARI Sharisha 11) seedlings were subjected to two different levels of drought with 10 and 20% polyethylene glycol (PEG) for 48 h. Leaf relative water content (RWC), chl *b* and chl (*a* + *b*) decreased but Pro content increased. Disrupting the antioxidant defense system, drought stress increased oxidative damage which was indicated by high MDA and H_2O_2 levels. Supplementation of SA in drought-stressed seedlings increased the leaf RWC and chl content, increased the AsA and GSH, decreased the GSSG content, and maintained a higher ratio of GSH/GSSG. Salicylic acid increased the activities of monodehydroascorbate reductase (MDHAR), DHAR, GR, glutathione peroxidase (GPX), CAT, glyoxalase I (Gly I), and glyoxalase II (Gly II) in drought affected seedlings as compared to the drought-stressed plants without SA supplementation, with a concomitant decrease in H_2O_2 and lipid peroxidation level [44]. Methyl-SA (at 0.1 mM) spray promoted drought-induced leaf senescence in *Salvia officinalis* [75]. Drought stress adversely affected growth performance of winter wheat, Cheyenne. Application of SA analogue 4-hydroxybenzoic acid (4-HBA) increased drought tolerance of winter wheat Cheyenne [16]. Foliar application of SA (10 μM) protects lemongrass (*Cymbopogon flexuosus* Steud. Wats.) varieties (Neema and Krishna) from drought stress by improving growth parameters, modulating the activities of nitrate reductase, carbonic anhydrase, and EL, Pro content, free amino acid, and in PEP carboxylase activity [76].

3.3. Extreme temperatures

Temperature is one of the vital factors that determine plants establishment, growth, development, and productivity. Due to climate change, global average temperature is fluctuating very rapidly and threatening the survival of living beings. Thus, among the various abiotic stresses, extreme temperature has become the talk of the topic in recent decades because of its devastating and damaging effects on plants [84]. Extreme temperature includes both high temperature (HT) and low temperature (LT) that can injure plants. The high temperature is the increasing of temperature beyond the critical threshold level that can deplete plant growth and metabolism depending on the sufficient time period [85]. Heat stress often becomes worse because of its combination with other stresses including drought [86]. High temperature severely alters the plant physiological processes including germination, photosynthesis, respiration, transpiration, partitioning of dry matter, etc. [87]. In addition, HT results in enzyme inactivation, protein denaturation, disruption of proteins and membranes which ultimately affects plant growth [88, 89]. Low-temperature consists of both freezing (<0°C) and chilling (0–15°C) temperatures. In chilling stress plant faces injury without formation of ice whereas, in freezing stress, the formation of ice occurs in plant tissues. Chilling and freezing stresses are together called cold stress or LT stress. Low-temperature stress shows various damaging symptoms in plants including faster senescence and decay [90, 91], interference with germination, cell membrane disruption, photosynthesis, water and nutrients uptake, reproductive development as well as growth and yield [92]. Either HT or LT conditions, at molecular level, leads to the overproduction of ROS which ultimately gives rise to the oxidative stress [84, 93]. Nowadays, to develop temperature-stress tolerance, the use of exogenous SA is one of the common approaches. Salicylic acid being the endogenous growth regulator or phytohormone acts as an important signaling molecule and develops abiotic stress tolerance in plants [94]. Recent advances on SA-mediated temperature stress tolerance have been listed in **Table 3**.

High temperature (30°C) resulted in a significant reduction in FW and soluble starch synthase activity of *T. aestivum* [46]. Foliar application of SA improved the FW, total RNA, and soluble starch synthase activity. In *Z. mays*, HT (40 ± 1°C) induced oxidative damage and reductions in dry biomass were reversed by exogenous SA treatment. SA developed HT tolerance by improving CAT, SOD, and POX activities [48]. The effects of SA on seed germination and physiological attributes of heat stressed (32/26°C, 12/12 h, day/night) *L. lycopersicum* were investigated. It has been revealed that SA reduced the germination time and increased the germination percentage together with the increased vitamin, lycopene, total soluble solid (TSS) and titratable acidity (TA) contents [49]. Temperature above 40°C, increased TBARS and H_2O_2 contents but decreased the net photosynthesis, RuBisCo activity, chl, and WUE of *T. aestivum* plant. Negative HT effects were counteracted significantly by exogenous SA supplementation [13]. Heat stress (45°C) in *Digitalis trojana Ivanina,* compared to its normal temperature, lowered the important antioxidant enzymes (CAT and SOD) activities. Pretreatment with SA significantly increased CAT and SOD activities with increased Pro, phenolic, and flavonoid contents [47]. The sharp decline of photosynthetic apparatus was found in *Vitis vinifera* in response to heat (43°C) stress. Alleviation of photosynthetic rate and RuBisCo activity were found when pretreated with SA [95]. *Cannabis sativa* induced thermotolerance against (40°C) temperature, when supplied with exogenous SA, were studied. Improved activities of antioxidant enzymes (SOD, DHAR, GPX, APX, and GR) were documented with decreased CAT activity. Decreased EL percentage with reduced H_2O_2 and TBARS contents were also evident [97]. Salicylic acid involved in various protective functions in *B. juncea* after HT (30 or 40°C) exposure. However, increased growth, g_s and CO_2 fixation along with improved defense system were found with SA treatment [98]. Chilling (5°C) stressed *Musa acuminate* when treated with exogenous SA increased SOD, CAT, and APX activities with decreased H_2O_2 accumulation [99]. Low temperature (3°C) disrupted the RuBPC and PEPC activities with decreased rate of CO_2 assimilation and respiration. Treatment with SA improved the activity and ameliorated the chilling effects [100]. Performance of *V. vinifera* was investigated upon HT (38 – 0.5°C) [101]. Pretreatment with SA increased H^+ and Ca^{2+}-ATPase activities with cerium phosphate grain appearance and thus gave higher stress tolerance. Wang and Li [102] noted that upon freezing (−3°C) stress in *V. vinifera,* besides upregulating antioxidants, increased maintenance of AsA-GSH pool and cytosolic Ca^{2+} homeostasis caused improved heat stress tolerance. High-temperature stress mediated increased ROS generation and oxidative stresses have been reported in several other plant species. Exogenous SA treatment resulted in the reduced ROS generation and oxidative stress in *Pratylenchus pratensis* [103], *Arabidopsis thaliana* [109]. Night ambient temperature ranges from 27 to 32°C in *Oryza sativa* caused the significant reduction in spikelet fertility and grain size. Exogenous SA treatment improved the rice grain fertility and hence, increased yield [104]. *Prunus persica* fruits were pretreated with SA before imposition of chilling injury (0°C). Reduced chilling injury was observed due to higher activities of antioxidants and heat shock protein 101 (HSP101) expression [105]. Leaves of heat (38/30°C, day/night) stressed *Rhododendron* became withered, defoliated, and brown. Total soluble protein and Chl contents were also reduced. Lower damage rate of leaves with higher chl and soluble protein were observed when supplemented with SA [106]. In a study with HT stressed (50°C) *V. radiata,* SA treatment increased the CAT, APX, POD, and GSH contents with enhanced defense system and radical scavenging mechanism [107].

3.4. Toxic metal/metalloids

In the industrial era, the most important and potential threat for crop production is the abiotic stress. Among them, toxic metal stress is one of the major concerns. Growing population and fast industrialization coincide together, results in the generation and dissemination of huge amount of toxic metals in the environment [110]. Toxic metal consists of a set of harmful elements having no biological role in organisms such as Cd, Pb, Hg, St, Al, etc. Although toxic metals and heavy metals (HMs) are often thought to be synonymous, some lighter metals such as Al may also cause toxicity. Toxic and HMs are differed in the case of their biological role.

Some HMs having a biological role in plants also considered toxic when they are used in high concentrations, viz. Ni, Cu, Zn, etc. On the other hand, metalloid includes those elements that show behavior both like metals and nonmetals including B, Si, Ge, St, As, Sb, etc. The underlying parent material and atmosphere are the two main sources of toxic metals. Metals are uptaken and accumulated easily by plants and causes toxicity within the plant tissue. They directly interact with the proteins, enzymes, and causes phytotoxicity. The inhibition of growth rate is the most certain consequences of metal toxicity [111]. Leaf rolling, chlorosis, necrosis, stunted growth, stomatal dysfunctioning, cation efflux, reduced water potential, alterations in the membrane, photosynthesis, metabolism, and various key enzymes are some other toxic metal effects in plants [111, 112]. Toxic metals also manipulate the nutrient homeostasis, water uptake, transport, transpiration, respiration, and ultimately may lead to plant death [113, 114]. Metal toxicity at the cellular level results in the overproduction of ROS [110]. To mitigate metal induced stresses in plants, plant biologists are trying to develop new strategies. Salicylic acid is a very important molecule that induces defense responses against various toxic metal/metalloids stresses (**Table 4**).

Several research findings demonstrated that exogenously applied SA improved the growth and photosynthetic traits in different plants by reducing the damaging effects of toxic metals. *O. sativa* exposed to As (25 and 50 μM) [50], *A. thaliana* exposed to As (100 μM) [115], *T. aestivum* exposed to Cd (500 and 1000 μM) [121], and *Z. mays* exposed to Cr (500 ppm) [52] grown well under SA supplementation. In a recent study, SA pretreatment reduced the oxidative stress in *T. aestivum* after Cd (0.01, 0.1, and 1 mM) exposure. Cd stress increased the lipid peroxidation and EL percentage. But the exogenous application of SA significantly declined the MDA content and EL percentage [53]. Salicylic acid also evidenced to alleviate the oxidative stress induced by metal toxicity in several other plant species by decreasing the toxic effects of overproduced ROS and lipid peroxidation. In *O. sativa*, enhanced MDA and H_2O_2 contents induced by As (25 and 50 μM) were reduced by the exogenous SA pretreatment [50]. Similarly, adverse oxidative stress was also demonstrated in *Lolium perenne* as induced by Cd (100 μM). Results showed that increased accumulation of $O_2^{\bullet -}$, H_2O_2, and higher MDA were decreased by SA application. Some other research findings also supported that the SA mitigates metal-induced oxidative damage in *Pisum sativum* [136], *Brassica oleracea* var. botrytis [135], *Medicago sativa* [134], etc. It was reported that treatment with SA alleviated the Cd (15 μM) induced root growth inhibition and improved the antioxidant activities thus reduced the Cd-induced oxidative stress in *Hordeum vulgare* [51]. Salicylic acid supplementation in As (100 μM) stressed *A. thaliana* showed improved performance in terms of antioxidant enzymes (APX, CAT, SOD) activities and enhanced tolerance to metal stress. In other experiment it was demonstrated that when SA was exogenously applied against Cd (0.25, or 0.50 mM) stress, increased amelioration of metal stress was observed with increasing activities of defense responsive genes and upregulating the antioxidant (SOD, CAT, APX, and GR) enzymes [117]. Decreased root growth, RWC, and increased oxidative stress were decreased by seed priming with SA [118]. Under Cd (50 and 100 mM) stress, plant growth, chl, total lipid contents, and nutrient absorption became decreased which were further increased by soaking seeds with SA. Increased stress tolerance with reduction of oxidative damage was also evident [119]. Increased ROS production, TBARS content, and EL with increased SOD, CAT, and APX activities were found after Cd (50 and 100 μM) exposure.

Seed priming with SA significantly decreased the oxidative damage and increased the antioxidant enzymes activities. Agami and Mohamed [121] reported that SA efficiently alleviated the adverse Cd (500 and 1000 μM) stress by restoring the growth parameters and increasing the antioxidant defense system. Exogenous SA developed Cr (mg L^{-1}) stress tolerance by improving growth, photosynthetic pigments and oxidative stress reduction by upregulating antioxidant defense system [52]. Inhibited antioxidant enzymes (APX, CAT, POX, and SOD), soluble sugars and chl contents were showed when *T. aestivum* exposed to Pb (10, 50, 100, 200 mg L^{-1}). Alleviated inhibitory effects were found after SA supplementation as cotreatment [126]. In *Zygophyllum fabago*, increased Pb accumulation after Pb (0.75 mM) exposure was reduced by SA pretreatment. Upregulated antioxidants and downregulated oxidative damage together induced stress tolerance [125]. Salicylic acid treatment against Cd (5, 10 or 50 mM) stress increased tolerance in *P. pratensis* by controlling uncontrolled absorption of Cd and maintaining nutrients (K, Ca, Fe, Mg) homeostasis [127]. Effects of B was investigated (25 and 50 mg kg^{-1}) toxity in *Daucus carota* and *S. oleracea*. In both plants, the growth, physiology, and antioxidant enzymes activity were affected by B toxicity. But exogenous SA application showed some protective effects in the alleviation of the metal toxicity [128, 129]. A similar finding was also demonstrated by Shi and Zhu [131] in *Crocus sativus* plant. Exposure to a toxic level of Mn (600 μM) with SA in *C. sativus* plant, maintained nutrient (Ca, Mg, Zn) homeostasis, reduced metal stress, and improved tolerance. Recent findings in *B. juncea* against Mn (3, 6, or 9 mM) toxicity revealed that SA is an important regulator of photosynthetic enzymes including carbonic anhydrase (CA). It together with the upregulated defense system and reduced oxidative damage-regulated the photosynthesis in a concentration-dependent manner [130]. Besides upregulation of antioxidant enzymes, SA involved in the activation of CaM-like protein genes and cytosolic Ca^{2+} in Al (30 μM) stressed *Glycine max*. It has been showed that increased metal stress tolerance resulted from exogenous SA treatment [132]. Effect of exogenous SA was investigated upon Pb (50 μM) stressed *Vallisneria natans*. Decreased chl, carotenoid, NADPH oxidase, nonprotein thiols, AsA but increased CAT, DHAR, and POD activities were observed. Pretreatment with SA increased the activities of NADH oxidase, AsA, and GSH in Hg (10 μM) stressed *M. sativa* and thus developed better tolerance to stress [133]. Recent evidences suggested that SA develops stress tolerance by involving in the regulation of photosynthetic pigments, activities of CA, NR, and anticancer alkaloids (vincristine and vinblastine) upon Ni (50, 100, and 150 mg kg^{-1}) exposure in *Catharanth roseus* [122]. Toxic level of Ni (0.5 mM) impacts on *Brasicca napus* also suggested that application of SA decreased the leaves' toxicity symptoms (chlorosis, necrosis, etc.) and improved growth and survival [123]. Recently a combined effect of metals (Co, Ni, Cd, Cr, and Pb) (0.25 M) on *B. oleracea* var. botrytis has been investigated. It has been found that SA alleviated all the toxic effects of metals except for Cr. In the case of Cr, SA accelerated the toxic effects and the plant died [135]. Improved CO_2 fixation, RuBPC activity, and chl content were found in Cd (0.5, 1, 2, and 5 μM) stressed *P. sativum* when supplied with exogenous SA.

3.5. Ozone and ultraviolet radiation

Due to the gradual increase in atmospheric ozone (O_3) concentration, it has become a major threat for plant species mainly because of its pollutant and photochemical oxidant affects [139]. Significant crop losses due to O_3 damage is predicted to be increased by 25% in background O_3

concentration over the next 30–50 years [140]. High concentrations of ozone induce oxidative stress, which activates programmed cell death and significantly inhibits plant growth, causing plant death and loss of quality [141]. It is the most noteworthy atmospheric pollutant in terms of phytotoxicity. It is to be noted that in the concentration of O_3 has been decreased by 5% in the past 50 years due to the release of anthropogenic pollutants and, a larger proportion of the UV radiation (especially (UV-B) spectrum reaches the Earth's surface [142]. Although sunlight plays an integral role in harvesting light energy through photosynthesis, high light, especially ultraviolet (UV) radiation, resulted in stress to plants, which include damage to DNA, proteins, and other cellular components [143]. This episode is unavoidable as 7% of the electromagnetic radiation emitted from the sun is in the UV range (200–400 nm). UV radiation also leads to oxidative stress by photooxidation and excess generation of ROS [84]. To cope up the adverse effects of both O_3 and UV radiation needs some adaptive mechanisms. In few plant species, SA was found to take part in enhancing the tolerance to O_3 and UV radiation mainly by enhancing antioxidant defense and improving plant growth.

In *A. thaliana,* UV-C light stress activated the transition to flowering through SA. SA could regulate the time of flowering by inducing photoperiod and autonomous pathways which are evident by late flowering in SA-deficient plants. While investigating the genes responsible for flowering induction viz. *constans (CO), flowering locus T (FT), suppressor of overexpression of constans 1 (SOC1),* and *flowering locus C (FLC)* it was observed that the expression of *CO, FT,* and *SOC1* transcripts decreased to around 50% in long day-grown SA-deficient plants when compared to contron plants [54]. In short day plants, only the levels of *FT* transcripts were reduced compared to *CO.* Thus, it indicated that SA might play important role in flowering under UV radiation [54]. The effect of UV was investigated (UV-A: 320–390 nm, UV-B: 312 nm, and UV-C: 254 nm radiation with a density of 6.1, 5.8, and 5.7 W m^{-2}) on *Capsicum annuum* plants and found that activities of antioxidant enzymes were enhanced in leaves in response to UV-B and UV-C radiation. Moreover, SA treatment showed further enhancement in the activities of POD, APX, CAT, and GR while some other enzymes were modulated [55]. In another report, a clear decline was reported in photosynthetic pigments (chl *a*, chl *b*, and carotenoid) under UV-A, UV-B, and UV-C, while a foliar spray of SA recovered this decline. The level of anthocyanins, flavonoids, and rutin in SA-treated plants was also higher than in a UV-exposed plant grown without SA [56]. *V. radiata,* exposed to UV-B radiation (ambient+4–8kJ m^{-2}) showed declined growth, photosynthetic pigments and photosynthesis (F_v/F_m and qP except NPQ) which were accompanied by significant decrease in SA level [57]. UV radiation also causes overproduction of ROS and concomitantly damaging effects on lipids, proteins, and membrane stability. However, SA pretreatment significantly alleviated the adverse effects. They also revealed that UV-B altered SA biosynthesis and SA-pretreatment might act as a signal that reduces oxidative stress by triggering upregulation of antioxidant defense and subsequent improvement of growth and photosynthesis [57]. In *Satureja hortensis,* both UV-B and UV-C exhibited decreased plant growth (plant height, root length, shoot DW, and leaf area), node number, internode distance and chl content, while stem diameter, leaf thickness, flavonoid content, phenolic content, and antioxidant activity were increased [58]. The increase in secondary metabolite such as flavonoid content, phenolics might be able to protect cells against free radicals but this level was not well enough under severe stress. On the other hand, plants treated with 1 mM SA exhibited higher growth and improved physiology compared to nontreated one and subsequently showed better appearance under UV radiation [58].

4. Salicylic acid and ROS detoxification

Salicylic acid is the most studied phytohormone regarding its role in oxidative stress. Due to its multifarious actions, it has been found very effective in detoxifying ROS in plant cells (**Figure 2**). According to Janda and Ruelland [144], SA-induced tolerance to abiotic stresses such as chilling, heat, heavy metals, osmotic stress, and salinity is involved in activation of the stress-induced antioxidant system. It has been demonstrated that SA could significantly improve both photosynthesis parameters and antioxidant defense system in conferring salt stress tolerance in *V. radiata* [17]. In *O. sativa*, exogenous SA significantly reduced the oxidative burst by reducing H_2O_2 and $O_2^{\bullet -}$ contents under herbicide exposure which was mainly due to the SA-mediated upregulation of antioxidant defense enzymes (SOD, POD, CAT, APX, GR, and GST) and efficient GSH pool. The positive role of SA is mostly dependent on their dose and application methods [145]. Other study also proved that SA-mediated antioxidative defense system was dependent on the concentration used and the method of application [146]. In *T. aestivum*, 0.25 mM SA resulted in marked increase in the antioxidant enzyme activities (SOD, CAT, POD, GPX, APX, and GR), while the treatment with 2.5 mM SA resulted in a decrease in the activities under water stress [146]. A lower dose of SA also maintains higher AsA pool which in turns significantly scavenged the ROS. This ROS detoxification induced by SA also associated with improved photosystem II (PSII) efficiency. Belkadhi et al. [147] showed that *Linum usitatissimum* plants showed a lower amount of lipid and protein oxidation and membrane oxidation under Cd stress when pretreated with SA. This protection was availed by the enhanced activities of SOD, GPX, and APX as well as the 2,2′-diphenyl-1-picrylhydrazyl (DPPH) scavenging activity and ferric-reducing antioxidant power (FRAP). Increases in MDHAR, DHAR, GR, GPX, and CAT activities 53, 64, 49, 82, and 65% were noticed in PEG-treated *B. juncea* seedlings when sprayed with 50 μM SA [44]. SA-induced upregulation of antioxidant enzymes caused 32 and 26% decrease in MDA and H_2O_2 content compared to drought (20% PEG, 48 h) alone. One of our research results showed that exogenous SA enhanced the activities of antioxidant enzymes and nonenzymatic antioxidants under salt stress (200 mM NaCl, 48 h). Compared to salt stress alone, NaCl + SA resulted in 41, 107, 25, 37, 44, and 59% increases in MDHAR, DHAR, GR, GST, GPX, and CAT activities [40]. Salicylic acid supplementation also increased AsA and GSH content by 48 and 39%, respectively and enhanced GSH/GSSG ratio by 47% compared to salt stress alone. As a result, MDA and H_2O_2 contents decreased by 39 and 31% [40]. In *Nitraria tangutorum* SA mitigated salt-induced oxidative stress (evidenced by a marked reduction in MDA and H_2O_2 content) by upregulating the activities of SOD, POD, and CAT. The content of MDA in the 1.5 mM SA treated seedlings under 100–400 mM NaCl treatments declined to 2.27–3.59 fold of the control which was a clear sign of the reduction of oxidative stress [148]. However, some of the enzymes like APX activity was inhibited at higher concentrations (1.0 and 1.5 mM) of SAMDA content was measured with 1.5 mM SA applied, and the contents of MDA in the leaves of SA-treated seedlings under 100–400 mM NaCl treatments declined to only 2.27–3.59 fold of the control.

http://dx.doi.org/10.5772/intechopen.68213

5. Interaction of SA with other signaling molecules

Salicylic acid not only exerts its positive effect independently but also interacts with other signaling molecules, phytohormones, and other phytoprotectants. These interactions show different signaling events and ultimate protection to plants from stress-induced damages (**Figure 3**). In grapevines, Wang and Li [102] showed improved Ca^{2+} homeostasis and associated antioxidant defenses under heat and cold stress-regulated by SA. When plants were treated with exogenous SA, they showed enhanced PM-Ca^{2+}ATPases and V-Ca^{2+} ATPases activities. Moreover, Ca^{2+} precipitates were shown on the inner side of the plasma membrane, and less were in intercellular spaces and the vacuole. However, in SA treated plants Ca^{2+} precipitates were in vacuoles, and few were on the inner side of the plasma membrane. Ca^{2+} precipitates in chloroplasts were bigger even after heat or cold stress. Importantly, SA treatment caused enhancement of the activities of APX, GR, and MDHAR, which efficiently reduced the lipid peroxidation and relative EL (REL), which concluded that exogenous SA could mitigate oxidative stress by maintaining Ca^{2+} homeostasis under extreme temperature stress [102]. In *O. sativa,* Wang et al. [145] reported that SA treatment downregulated ABA genes more in cultivar XS 134, which correlated with the enhanced tolerance to quinclorac-induced oxidative stress. Application of SA had obvious effects on all of the ABA-related genes and inhibited the expression of *OsABA8ox1, OsABA8ox2, OsABA8ox3, OsNCED1, OsNCED2,* and *OsNCED3* as compared to quinclorac stress alone. Since overproduction of ABA and ROS is highly associated this downregulation protected the plants from herbicide-induced damages. Wang et al. [145] also reported SA-induced inhibition of ABA synthesizing enzymes. Leslie and Romani [149] reported that SA inhibited ethylene formation which triggered biosynthesis of ABA under stress conditions [150]. In the adventitious roots of *Panax ginseng,* SA-induced enhancement of the activities of NADPH oxidase, SOD, CAT, POD, and APX was evident while no significant effect on AsA and GSH content were observed [151]. These effects were mostly NO-dependent and it was also observed that SA-induced the generation of NO. They revealed that at lower concentration (100 μM) SA was highly effective in inducing the

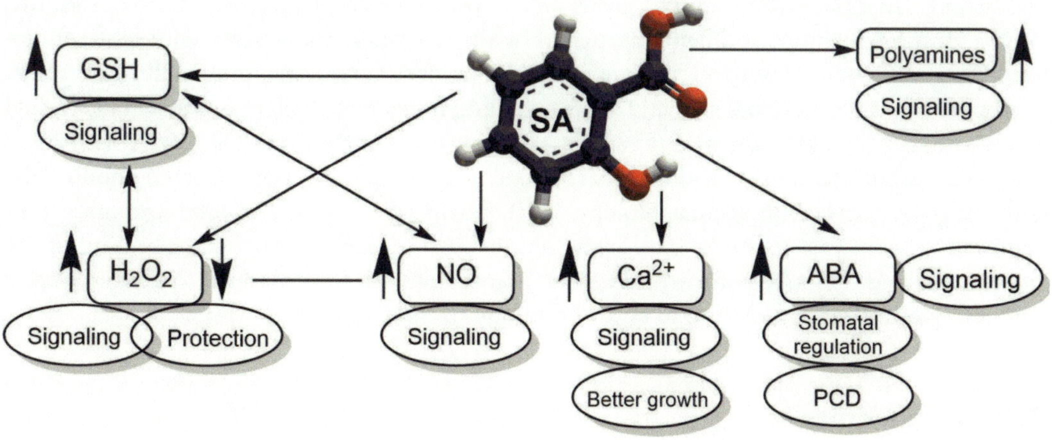

Figure 3. Interaction of SA with other signaling molecules to elicit defense responses in plants.

accumulation of NO, $O_2^{\bullet -}$ and it took part in stress signaling. Interactive effects of SA and NO were studied in mitigating osmotic stress (−0.4 MPa) in *T. aestivum*. It was observed that osmotic stress induced chl degradation and membrane instability, and H_2O_2 generation and lipid peroxidation were effectively reduced by exogenous application of SA or SNP, which was associated with the enhancement of antioxidant defense. However, pretreatment of plants with methylene blue (MB; as a guanylate cyclase inhibitor) reversed or reduced the protective effects of SA and SNP suggesting that the protective effects were likely attributed to NO signaling. They also concluded that NO may act as downstream of SA signaling in the reduction of induced oxidative damage [152]. SA-mediated H_2O_2 signaling and subsequent Cd stress tolerance was revealed in *L. usitatissimum* [147]. Seedlings pretreated with 250 or 1000 μM SA resulted in enhanced production of H_2O_2 because of inhibited CAT activity. Although the control plants with SA pretreatment showed significant (1.2 fold) increase in H_2O_2, this level is remarkably lower when compared with Cd alone and Cd+SA. These results indicated that SA could regulate the Cd-induced oxidative stress because Cd-treated seedlings primed with SA exhibited a higher level of total antioxidant capacities and increased activities of H_2O_2-detoxifying enzymes [147]. Exogenous SA application was found to activate GSH synthesis in *B. juncea* and *B. napus* and showed enhanced protection against drought- and salt-induced oxidative damages [44, 40].

6. Conclusions and perspectives

Salicylic acid plays an important role in the regulation of growth and physiology in relation to the abiotic stress responses of plants. The SA was found to be effective in the different form of application foliar spray/incorporation with growing media depending upon plant species. The low concentration of SA showed advantageous effects in abiotic stress tolerance of plants. In contrast, the high concentration of SA showed toxic effects. Thus, both the concentration and application method of SA are critical to obtaining its best effect on different plant species. In the biosynthesis pathway of SA, there are unknown steps and enzymes which should be discovered. The catabolism and the further fate of transformed product of SA are not known clearly. How SA interacts and being regulated by the cross-talk in harmony with other phytohormones and plant growth regulators working (auxins, cytokinins, gibberellins, ethylene, jasmonates, brassinosteroids, etc.) and other signaling molecules (NO, H_2O_2) were not studied extensively. SA-mediated defense networks and insights into the cross-talk of SA with other defense-signaling pathways should be revealed. An integrated approach combining the knowledge of genetics, molecular biology, biochemistry, genomics, and bioinformatics techniques is a useful tool to study the functioning of SA in plants. Clear understanding of the biosynthesis and catabolic pathway and other unanswered question are vital to exploit SA as a potent phytoprotectant molecule to improve abiotic stress tolerances.

Acknowledgements

We are highly thankful to Mazhar Ul Alam, Laboratory of Plant Stress Responses, Faculty of Agriculture, Kagawa University, Japan, for his critical reading and formatting of the manuscript draft. The first author acknowledges Japan Society for the Promotion of Science (JSPS) for funding in his research. We are also highly thankful to Mr. Md. Mosfeq-Ul-Hasan, Zhejiang University, Hangzhou, China, for providing us several supporting articles. As page limitation precluded us from citing a large number of studies, we apologize to those whose original publications are therefore not directly referenced in this chapter.

Author contributions

M.H. performed literature reviews and drafted the manuscript; K.N., T.I.A. and T.F.B., M.I. and M.H. contributed the review for literature research; M.F. and H.O. reviewed the manuscript and approved the final draft.

Conflicts of interest

The authors declare no conflict of interest.

Abbreviations

4-HBA	4-hydroxybenzoic acid
ABA	Abscisic acid
APX	Ascorbate peroxiddase
AsA	Ascorbate/ascorbic acid
BA2H	Benzoicacid-2-hydroxylase
CAT	Catalase
chl	Chlorophyll
C_i	Intercellular CO_2 concentration
DAS	Days after sowing
DHA	Dehydroascorbate
DHAR	Dehydroascorbate reductase
DPPH	2,2′-diphenyl-1-picrylhydrazyl
DW	Dry weight
EL	Electrolyte leakage
FRAP	Ferric reducing antioxidant power
FW	Fresh weight
GB	Glycinebetaine
Gly I	Glyoxalase I
Gly II	Glyoxalase II
GO	Glycolate oxidase
GR	Glutathione reductase
g_s	Substomatal conductance
GSH	Reduced glutathione
GSSG	Oxidized glutathione
GST	Glutathione *S*-tranferase

HSPs	Heat shock proteins
HT	High temperature
IC	Isochorismate
ICS	Isochorismate synthase
IPL	Isochorismate pyruvate lyase
LOX	Lipoxygenase
LT	Low temperature
MAPK	Mitogen-activated protein kinase
MB	Methylene blue
MDA	Malondialdehyde
MDHAR	Monodehydroascorbate reductase
MeSA	Methyl salicylate
MSI	Membrane stability index
NADH	Nicotinamide adenine dinucleotide phosphate
NO	Nitric oxide
PAL	Phenylalanine ammonia-lyase
PEG	Polyethylene glycol
P_n	Net photosynthesis
POD	Peroxidase
POX	Peroxidases
Pro	Proline
ROS	Reactive oxygen species
RWC	Relative water content
SA	Salicylic acid
SA-Asp	Salicyloyl-L-aspartic acid
SAG	Salicylic acid 2-O-β-glucoside
SAT	Serine acetyl transferase
SGE	Salicyloyl glucose ester
SNP	Sodium nitroprusside
SOD	Superoxide dismutase
TBARS	Thiobarbituric acid reactive substances
T_r	Transpiration rate
TSS	Total soluble solid
WUE	Water use efficiency

Author details

Mirza Hasanuzzaman[1,2], Kamrun Nahar[3], Tasnim Farha Bhuiyan[3,4], Taufika Islam Anee[1,4], Masashi Inafuku[2], Hirosuke Oku[2] and Masayuki Fujita[4]*

*Address all correspondence to: fujita@ag.kagawa-u.ac.jp

1 Department of Agronomy, Faculty of Agriculture, Sher-e-Bangla Agricultural University, Dhaka, Bangladesh

2 Molecular Biotechnology Group, Center of Molecular Biosciences (COMB), Tropical Biosphere Research Center, University of the Ryukyus, Nishihara, Okinawa, Japan

3 Department of Agriculture Botany, Faculty of Agriculture, Sher-e-Bangla Agricultural University, Dhaka, Bangladesh

4 Laboratory of Plant Stress Responses, Department of Applied Biological Science, Faculty of Agriculture, Kagawa University, Miki-cho, Kagawa, Japan

References

[1] Khan MIR, Khan NA. Salicylic acid and jasmonates: Approaches in abiotic stress. Journal of Plant Biochemistry and Physiology. 2013;**1**:e113

[2] Anjum NA, Gill SS, Gill R. (eds) Plant adaptation to environmental change: significance of amino acids and their derivatives. Wallingford: CABI; 2014. doi: 10.1079/9781780642734.0000

[3] Kaur G, Kumar S, Nayyar H, Upadhyaya HD. Cold stress injury during the pod-filling phase in Chickpea (*Cicer arietinum* L.): Effects on quantitative and qualitative components of seeds. Journal of Agronomy and Crop Science 2008;**194**:457–464

[4] Mantri N, Patade V, Penna S, Ford R. Abiotic stress responses in plants: present and future. In: Ahmad P, Prasad MNV (eds) Abiotic Stress Responses in Plants. New York: Springer; 2012. pp. 1–19

[5] Raskin I. Role of salicylic acid in plants. Annual Review of Plant Physiology and Plant Molecular Biology. 1992;**43**:439–463

[6] Yalpani N, Enyedi AJ, Leon J, Raskin I. Ultraviolet light and ozone stimulate accumulation of salicylic acid, pathogenesis-related proteins and virus resistance in tobacco. Planta. 1994;**193**:372–376

[7] Senaratna T, Touchell D, Bunn E, Dixon K. Acetyl salicylic acid (aspirin) and salicylic acid induce multiple stress tolerance in bean and tomato plants. Plant Growth Regulation. 2000;**30**:157–161

[8] Klessig DF, Malamy J. The salicylic acid signal in plants. Plant Molecular Biology. 1994;**26**:1439–1458

[9] Harper JR, Balke NE. Characterization of the inhibition of K^+ absorption in oats roots by salicylic acid. Plant Physiology. 1981;**68**:1349–1353

[10] Khan W, Prithviraj B, Smith DL. Photosynthetic responses of corn and soybean to foliar application of salicylates. Journal of Plant Physiology. 2003;**160**:485–492

[11] Shirasu K, Nakajima A, Rajshekar K, Dixon RA, Lamb C. Salicylic acid potentiates an agonist-dependent gain control that amplifies pathogen signal in the activation of defence mechanism. Plant Cell. 1997; **9**:261–270.

[12] Nazar R, Iqbal N, Sayeed S, Khan NA. Salicylic acid alleviates decreases in photosynthesis under salt stress by enhancing nitrogen and sulfur assimilation and antioxidant metabolism differentially in two mungbean cultivars. Journal of Plant Physiology. 2011;**168**:807–815

[13] Khan MIR, Iqbal N, Masood A, Per TS, Khan NA. Salicylic acid alleviates adverse effects of heat stress on photosynthesis through changes in proline production and ethylene formation. Plant Signaling & Behavior. 2013;**8**:e26374

[14] Miura K, TadaY. Regulation of water, salinity, and cold stress responses by salicylic acid. Frontiers in Plant Science. 2014;**5**:4. doi: 10.3389/fpls.2014.00004.

[15] Kumar D. Salicylic acid signaling in disease resistance. Plant Science. 2014;**228**:127–124.

[16] Horváth E, Pál M, Szalai G, Páldi E, Janda T. Exogenous 4-hydroxybenzoic acid and salicylic acid modulate the effect of short-term drought and freezing stress on wheat plants. Biologia Plantarum. 2007; **51**:480—487

[17] Khan MIR, Asgher M, Khan NA. Alleviation of salt induced photosynthesis and growth inhibition by salicylic acid involves glycinebetaine and ethylene in mungbean (*Vigna radiata* L.). Plant Physiology and Biochemistry. 2014;**80**:67–74

[18] Khan MIR, Khan NA. Ethylene reverses photosynthetic inhibition by nickel and zinc in mustard through changes in PSII activity, photosynthetic nitrogen use efficiency, and antioxidant metabolism. Protoplasma 2014;**251**:1007–1019

[19] Anwar S, Iqbal M, Raza SH, Iqbal N. Efficacy of seed preconditioning with salicylic and ascorbic acid in increasing vigor of rice (*Oryza sativa* L.) seedling. Pakistan Journal of Botany. 2013;**45**:157–162

[20] Palma F, López-Gómez M, Tejera NA, Lluch C. Salicylic acid improves the salinity tolerance of *Medicago sativa* in symbiosis with *Sinorhizobium meliloti* by preventing nitrogen fixation inhibition. Plant Science. 2013;**208**:75–82

[21] Chai J, Liu J, Zhou J, Xing D. Mitogen-activated protein kinase 6 regulates NPR1 gene expression and activation during leaf senescence induced by salicylic acid. Journal of Experimental Botany. 2014; **65**:6513–6528

[22] Wildermuth MC, Dewdney J, Wu G, Ausubel FM. Isochorismate synthase is required to synthesize salicylic acid for plant defence. Nature. 2001;**414**:562–565

[23] Uppalapati S R, Ishiga Y, Wangdi T, Kunkel BN, Anand A, Mysore KS, Bender CL. The phytotoxin coronatine contributes to pathogen fitness and is required for suppression of salicylic acid accumulation in tomato inoculated with *Pseudomonas syringae* pv. tomato *DC*3000. Molecular Plant-Microbe Interactions. 2007;**20**:955–965

[24] Catinot J, Buchala A, Abou-Mansour E, Metraux JP. Salicylic acid production in response to biotic and abiotic stress depends on isochorismatein *Nicotiana benthamiana*. FEBS Letters. 2008;**582**:473–478

[25] Dewdney J, Reuber TL, Wildermuth MC, Devoto A, Cui J, Stutius LM, Drummond EP, Ausubel FM. Three unique mutants of *Arabidopsis* identify edsloci required for limiting growth of abiotrophic fungal pathogen. The Plant Journal. 2000;**24**:205–218

[26] Garcion C, Lohmann A, Lamodiere E, Catinot J, Buchala A, Doermann P, Métraux J. Characterization and biological function of the ISOCHORIS-MATESYNTHASE2 gene of *Arabidopsis*. Plant Physiology 2008;**147**:1279–1287

[27] Wan D, Li R, Zou B, Zhang X, Cong J, Wang R, Xia Y, Li G. Calmodulin-binding protein CBP60g is a positive regulator of both disease resistance and drought tolerance in *Arabidopsis*. Plant Cell Reports 2012;**31**:1269–1281

[28] Mercado-Blanco J, VanDerDrift KM, Olsson PE, Thomas-Oates JE, Van Loon, LC, Bakker PA. Analysis of the pmsCEAB gene cluster involved in biosynthesis of salicylic acid and the siderophore pseudomonine in the biocontrol strain *Pseudomonas fluorescens* WCS374. Journal of Bacteriology. 2001;**183**:1909–1920

[29] Ellis BE, Amrhein N. The 'NIH-shift' during aromatic ortho-hydroxylation in higher plants. Phytochemistry. 1971;**10**:3069–3072

[30] Chadha KC, Brown SA. Biosynthesis of phenolic acids in tomato plants infected with *Agrobacterium tumefaciens*. Canadian Journal of Botany. 1974;**52**:2041–2047

[31] Yalpani N, Leon J, Lawton MA, Raskin I. Pathway of salicylic acid biosynthesis in healthy and virus-inoculated tobacco. Plant Physiology. 1993;**103**:315–321

[32] Song JT, Koo YJ, Seo HS, Kim MC, Choi YD, Kim JH. Overexpression of AtSGT1, an Arabidopsis salicylic acid glucosyltransferase, leads to increased susceptibility to *Pseudomonas syringae*. Phytochemistry 2008;**69**:1128–1134

[33] Chen F, D'Auria JC, Tholl D, Ross JR, Gershenzon J, Noel JP, Pichersky E. An *Arabidopsis thaliana* gene for methylsalicylate biosynthesis, identified by a biochemical genomics approach, has a role in defense. The Plant Journal. 2003;**36**:577–588

[34] Zhang Z, Li Q, Li Z, Staswick PE, Wang M, Zhu Y, He Z. Dual regulation role of GH3.5 in salicylic acid and auxin signaling during Arabidopsis-*Pseudomonas syringae* interaction. Plant Physiology. 2007;**145**:450–464

[35] Baek D, Pathange P, Chung JS, Jiang J, Gao L, Oikawa A, Hirai MY, Saito K, Pare PW, Shi H. A stress-inducible sulphotransferase sulphonates salicylic acid and confers pathogen resistance in Arabidopsis. Plant Cell and Environment. 2010;**33**:1383–1392

[36] Mustafa NR, Verpoorte R. Phenolic compounds in *Catharanthus roseus*. Phytochemistry Reviews. 2007; **6**:243–258

[37] Dempsey DA, Vlot AC, Wildermuth MC, Klessig DF. Salicylic acid biosynthesis and metabolism. Arabidopsis Book. 2011;**9**:e0156. DOI: 10.1199/tab.0156

[38] Szepesi A, Csiszára J, Gémesa K, Horvátha E, Horvátha F, Simonb ML, Tari I. Salicylic acid improves acclimation to salt stress by stimulating abscisic aldehyde oxidase activity and abscisic acid accumulation, and increases Na^+ content in leaves without toxicity symptoms in *Solanum lycopersicum* L. Journal of Plant Physiology. 2009;**166**:914–925

[39] Kumara GDK, Xia Y, Zhu Z, Basnayake BMVS, Beneragama CK. Effects of exogenous salicylic acid on antioxidative enzyme activities and physiological characteristics in gerbera (*Gerbera jamesonii* L.) grown under NaCl stress. Agriculture and Life Sciences. 2010;**36**:591–601

[40] Hasanuzzaman M, Alam MM, Nahar K, Mahmud JA, Ahamed KU, Fujita M. Exogenous salicylic acid alleviates salt stress-induced oxidative damage in *Brassica napus* by enhancing the antioxidant defense and glyoxalase systems. Australian Journal of Crop Science. 2014;**8**:631–639

[41] Nazar R, Umar S, Khan NA. Exogenous salicylic acid improves photosynthesis and growth through increase in ascorbate glutathione metabolism and S assimilation in mustard under salt stress. Plant Signaling & Behavior. 2015;**10**:e1003751

[42] Tari I, Csiszár J, Horváth E, Poór P, Takács Z, Szepesi A. The alleviation of the adverse effects of salt stress in the tomato plant by salicylic acid shows a time and organ-specific antioxidant response. Acta Biologica Cracoviensia Series Botanica. 2015;**57**:21–30

[43] Kang G, Li G, Xu W, Peng X, Han Q, Zhu Y. Proteomics reveals the effects of salicylic acid on growth and tolerance to subsequent drought stress in wheat. Journal of Proteome Research. 2012;**11**:6066–6079

[44] Alam MM, Hasanuzzaman M, Nahar K, Fujita M. Exogenous salicylic acid ameliorates short-term drought stress in mustard (*Brassica juncea* L.) seedlings by upregulating the antioxidant defense and glyoxalase system. Australian Journal of Crop Science. 2013;**7**:1053–1063

[45] Demiralay M, Sağlam A, Kadioğlu A. Salicylic acid delays leaf rolling by inducing antioxidant enzymes and modulating osmoprotectant content in *Ctenanthe setosa* under osmotic stress. Turkish Journal of Biology. 2013;**37**:49–59

[46] Kumar RR, Sharma SK, Goswami S, Verma P, Singh K, Dixit N, Pathak H, Viswanathan C, Rai RD. Salicylic acid alleviates the heat stress-induced oxidative damage of starch biosynthesis pathway by modulating the expression of heat-stable genes and proteins in wheat (*Triticum aestivum*). Acta Physiologiae Plantarum. 2015;**37**:1–12.

[47] Cingoz GS, Gurel E. Effects of salicylic acid on thermotolerance and cardenolide accumulation under high temperature stress in *Digitalis trojana Ivanina*. Plant Physiology and Biochemistry. 2016;**105**:145–149

[48] Khanna P, Kaur K, Gupta AK. Salicylic acid induces differential antioxidant response in spring maize under high temperature stress. Indian Journal of Experimental Biology. 2016;**54**:386–393

[49] Singh SK, Singh PK. Effect of seed priming of tomato with salicylic acid on growth, flowering, yield and fruit quality under high temperature stress conditions. International Journal of Advanced Research. 2016; **4**:723–727

[50] Singh AP, Dixit G, Mishra S, Dwivedi S, Tiwari M, Mallick S, Pandey V, Trivedi PK, Chakrabarty D, Tripathi RD. Salicylic acid modulates arsenic toxicity by reducing its root to shoot translocation in rice (*Oryza sativa* L.). Frontiers in Plant Science. 2015;**6**:340. doi: 10.3389/fpls.2015.00340.

[51] Tamás L, Mistrík I, Alemayehu A, Zelinová V, Boˇcová B, Huttová J. Salicylic acid alleviates cadmium-induced stress responses through the inhibition of Cd-induced auxin-mediated reactive oxygen species production in barley root tips. Journal of Plant Physiology. 2015;**173**:1–8

[52] Islam F, Yasmeen T, Arif MS, Riaz M, Shahzad SM, Imran Q, Ali I. Combined ability of chromium (Cr) tolerant plant growth promoting bacteria (PGPB) and salicylic acid (SA) in attenuation of chromium stress in maize plants. Plant Physiology and Biochemistry. 2016;**108**:456–467

[53] Shakirova FM, Allagulova ChR, Maslennikova DR, Klyuchnikova EO, Avalbaev AM, Bezrukova MV. Salicylic acid-induced protection against cadmium toxicity in wheat plants. Environmental and Experimental Botany. 2016;**122**:19–28

[54] Martínez C, Pons E, Prats G, León J. Salicylic acid regulates flowering time and links defence responses and reproductive development. Plant Journal. 2004;**37**:209–217

[55] Mahdavian K, Ghorbanli M, Kalantari KM. Role of salicylic acid in regulating ultraviolet radiation-induced oxidative stress in pepper leaves. Russian Journal of Plant Physiology. 2008;**55**:560. doi: 10.1134/S1021443708040195

[56] Mahdavian K, Kalantari KM, Ghorbanli M, Torkzade M. The effects of salicylic acid on pigment contents in ultraviolet radiation stressed pepper plants. Biologia Plantarum. 2008;**52**:170. doi: 10.1007/s10535-008-0037-0

[57] Singh VP, Kumar J, Singh S, Prasad SM. Dimethoate modifies enhanced UV-B effects on growth, photosynthesis and oxidative stress in mung bean (*Vigna radiata* L.) seedlings: Implication of salicylic acid. Pesticide Biochemistry and Physiology. 2014;**116**:13–23.

[58] Abrun A, Fattahi M, Hassani A, Avestan S. Salicylic acid and UV-B/C radiation effects on growth and physiological traits of *Satureja hortensis* L. Notulae Scientia Biologicae. 2016;**8**:170–175.

[59] Arfan M, Athar HR, Ashraf M. Does exogenous application of salicylic acid through the rooting medium modulate growth and photosynthetic capacity in two differently adapted spring wheat cultivars under salt stress? Journal of Plant Physiology. 2007;**164**:685−694

[60] Gunes A, Inal A, Alpaslan M, Eraslan F, Bagci EG, Cicek N. Salicylic acid induced changes on some physiological parameters symptomatic for oxidative stress and mineral nutrition in maize (*Zea mays* L.) grown under salinity. Journal of Plant Physiology. 2007;**164**:728−736

[61] Misra N, Saxena P. Effect of salicylic acid on proline metabolism in lentil grown under salinity stress. Plant Science. 2009;**177**:181−189

[62] Horváth E, Csiszár J, Gallé Á, Poór P, Szepesi Á, Tari I. Hardening with salicylic acid induces concentration-dependent changes in abscisic acid biosynthesis of tomato under salt stress. Journal of Plant Physiology. 2015;**183**:54–63

[63] Fayez KA, Bazaid SA. Improving drought and salinity tolerance in barley by application of salicylic acid and potassium nitrate. Journal of the Saudi Society of Agricultural Sciences. 2014;**23**:45−55

[64] Li T, Hu Y, Du X, Tang H, Shen C, Wu J. Salicylic acid alleviates the adverse effects of salt stress in *Torreya grandis* cv. Merrillii seedlings by activating photosynthesis and enhancing antioxidant systems. PLoS ONE. 2014;**9**:e109492

[65] Jayakannan M, Bose J, Babourina O, Rengel Z, Shabala S. Salicylic acid improves salinity tolerance in Arabidopsis by restoring membrane potential and preventing salt-induced K^+ loss via a GORK channel. Journal of Experimental Botany. 2013;**64**:2255–2268

[66] Li G, Peng X, Wei L, Kang G. Salicylic acid increases the contents of glutathione and ascorbate and temporally regulates the related gene expression in salt-stressed wheat seedlings. Gene. 2013; **529**:321−325

[67] Hasanuzzaman M, Nahar K, Gill SS, Fujita M. Drought stress responses in plants, oxidative stress, and antioxidant defense. In: Tuteja N, Gill SS (eds) Climate change and plant abiotic stress tolerance. Wiley: New York; 2014. pp. 209−249

[68] Kang GZ, Li GZ, Liu GQ, Xu W, Peng XQ, Wang CY. Exogenous salicylic acid enhances wheat drought tolerance by influence on the expression of genes related to ascorbate-glutathione cycle. Biologia Plantarum. 2013;**57**:718–724

[69] Miura K, Okamoto H, Okuma E, Shiba H, Kamada H, Hasegawa PM. SIZ1 deficiency causes reduced stomatal aperture and enhanced drought tolerance via controlling salicylic acid-induced accumulation of reactive oxygen species in *Arabidopsis*. Plant Journal. 2013;**49**:79–90

[70] Hayat S, Hasan SA, Fariduddin Q, Ahmad A. Growth of tomato (*Lycopersicon esculentum*) in response to salicylic acid under water stress. Journal of Plant Interactions. 2008;**3**:297−304

[71] Hussain M, Malik MA, Farooq M, Ashraf MY, Cheema MA. Improving drought tolerance by exogenous application of glycinebetaine and salicylic acid in sunflower. Journal of Agronomy and Crop Science 2008;**194**:193–199

[72] Korkmaz A, Uzunlu M, Demirkiran A. Treatment with acetyl salicylic acid protects muskmelon seedlings against drought stress. Acta Physiologiae Plantarum. 2007;**29**:503–508

[73] Khan SU, Bano A, Jalal-ud-din, Gurmani A. Abscisic acid and salicylic acid seed treatment as potent inducer of drought tolerance in wheat (*Triticum aestivum* L.). Pakistan Journal of Botany. 2012; **44**:43−49.

[74] Kadioglu A, Saruhan N, Sağlam A, Terzi R, Acet T. Exogenous salicylic acid alleviates effects of long term drought stress and delays leaf rolling by inducing antioxidant system. Plant Growth Regulation. 2011;**64**:27–37

[75] Abreu ME, Munné-Bosch S. Salicylic acid may be involved in the regulation of drought-induced leaf senescence in perennials: A case study in field-grown *Salvia officinalis* L. plants. Environmental and Experimental Botany. 2008;**64**:105–112

[76] Idrees M, Khan MMA, Aftab T, Naeem M, Hashmi N. Salicylic acid-induced physiological and biochemical changes in lemongrass varieties under water stress. Journal of Plant Interactions. 2010; **5**:293–303

[77] El Tayeb MA, Ahmed NL. Response of wheat cultivars to drought and salicylic acid. American-Eurasian Journal of Agronomy. 2010;**3**:1–7

[78] Yazdanpanah S, Baghizadeh A, Abbassi F. The interaction between drought stress and salicylic and ascorbic acids on some biochemical characteristics of *Satureja hortensis*. African Journal of Agricultural Research. 2011;**6**:798–807

[79] Rao SR, Qayyum A, Razzaq A, Ahmad M, Mahmood I, Sher A. Role of foliar application of salicylic acid and l-tryptophan in drought tolerance of maize. Journal of Animal and Plant Sciences. 2012;**22**:768–772.

[80] Bidabadi SS, Mahmood M, Baninasab B, Ghobadi C. Influence of salicylic acid on morphological and physiological responses of banana (*Musa acuminata* cv. 'Berangan', AAA) shoot tips to in vitro water stress induced by polyethylene glycol. Peeters Online Journals. 2012;**5**:33–39

[81] Habibi G. Exogenous salicylic acid alleviates oxidative damage of barley plants under drought stress. Acta Biologica Szegediensis. 2012;**56**:57–63

[82] Farooq MA, Wahid DJ, Lee SA, Cheema, Aziz T. Comparative time course action of the foliar applied glycinebetaine, salicylic acid, nitrous oxide, brassinosteroids and spermine in improving drought resistance of rice. Journal of Agronomy and Crop Science. 2010;**196**:336–345

[83] Azooz MM, Youssef MM. Evaluation of heat shock and salicylic acid treatments as inducers of drought stress tolerance in hassawi wheat. American Journal of Plant Physiology. 2010;**5**:56–70

[84] Hasanuzzaman M, Hossain MA, Teixeira da Silva JA, Fujita M. Plant responses and tolerance to abiotic oxidative stress: antioxidant defense is a key factor. In: Bandi V, Shanker AK, Shanker C, Mandapaka M (eds) Crop stress and its management: perspectives and strategies. Berlin: Springer; 2012. pp. 261–316

[85] Wahid A. Physiological implications of metabolites biosynthesis in net assimilation and heat stress tolerance of sugarcane (*Saccharum officinarum*) sprouts. Journal of Plant Research. 2007;**120**:219–228

[86] Mittler R. Abiotic stress, the field environment and stress combination. Trends in Plant Science. 2006; **11**:15–19

[87] Nahar K, Hasanuzzaman M, Ahamed KU, Hakeem KR, Ozturk M, Fujita M. Plant responses and tolerance to high temperature stress: role of exogenous phytoprotectants. In: Hakeem KR (ed) Crop Production and Global Environmental Issues. International Publishing Switzerland: Springer; 2015. pp. 385–435

[88] Howarth CJ. Genetic improvements of tolerance to high temperature. In: Ashraf M, Harris PJC (eds) Abiotic stresses: plant resistance through breeding and molecular approaches. New York: Howarth Press; 2005. pp. 277–300

[89] Hemantaranjan A, Bhanu AN, Singh MN, Yadav DK, Patel PK, Singh R, Katiyar D. Heat stress responses and thermotolerance. Advances in Plants & Agricultural Research. 2014;**1**:1–12

[90] Sharma P, Dubey RS. Drought induces oxidative stress and enhances the activities of antioxidant enzymes in growing rice seedlings. Plant Growth Regulation. 2005;**46**: 209–221

[91] Solanke AU, Sharma AK. Signal transduction during cold stress in plants. Physiology and Molecular Biology of Plants 2008;**14**: 69–79

[92] Hasanuzzaman M, Nahar K, Fujita M. Extreme temperature responses, oxidative stress and antioxidant defense in plants. In: Vahdati K, Leslie C (eds) Abiotic Stress – plant responses and applications in agriculture. Rijeka: InTech; 2013. doi: org/10.5772/54833.

[93] Suzuki N, Mittler R. Reactive oxygen species and temperature stresses: A delicate balance between signaling and destruction. Physiologia Plantarum. 2006;**126**:45–51

[94] Khan NA, Nazar R, Iqbal N, Anjum NA. Phytohormones and abiotic stress tolerance in plants. Berlin: Springer; 2012. doi: 10.1007/978-3-642-25829-9

[95] Wang L, Fan L, Loescher W, Duan W, Liu G, Cheng J, Luo H, Li S. Salicylic acid alleviates decreases in photosynthesis under heat stress and accelerates recovery in grapevine leaves. BMC Plant Biology. 2010;**10**:34. doi: 10.1186/1471-2229-10-34.

[96] Ghasemi M, Modarresi M, Jelodar NB, Bagheri N, Jamali A. The evaluation of exogenous application of salicylic acid on physiological characteristics, proline and essential oil content of chamomile (*Matricaria chamomila* L.) under normal and heat stress conditions. Agriculture. 2016;**6**:31. doi: 10.3390/agriculture6030031

[97] Shi Q, Bao Z, Zhu Z, Ying Q, Qian Q. Effects of different treatments of salicylic acid on heat tolerance, chlorophyll fluorescence, and antioxidant enzyme activity in seedlings of *Cucumis sativa* L. Plant Growth Regulation. 2006;**48**:127–135

[98] Hayat S, Masud A, Yusuf M, Fariduddin Q, Ahmad A. Growth of Indian mustard (*Brassica juncea* L.) in response to salicylic acid uder high-temperature stress. Brazilian Journal of Plant Physiology. 2009; **23**:187–195

[99] Kang G, Wang C, Sun G, Wang Z. Salicylic acid changes activities of H_2O_2 metabolizing enzymes and increases the chilling tolerance of banana seedlings. Environmental and Experimental Botany. 2003;**50**:9 –15.

[100] Yordanova R, Popova L. Effect of exogenous treatment with salicylic acid on photosynthetic activity and antioxidant capacity of chilled wheat plants. General and Applied Plant Physiology. 2007;**33**:155–170

[101] Liu Y, Zhang J, Liu H, Huang W. Salicylic acid or heat acclimation pre-treatment enhances the plasma membrane-associated ATPase activities in young grape plants under heat shock. Scientia Horticulturae 2008;**119**:21–27

[102] Wang L, Li S. Salicylic acid-induced heat or cold tolerance in relation to Ca^{2+} homeostasis and antioxidant systems in young grape plants. Plant Science. 2006;**170**:685–694

[103] He Y, Liu Y, Cao W, Huai M, Xu B, Huang B. Effects of salicylic acid on heat tolerance associated with antioxidant metabolism in kentucky bluegrass. Crop Science. 2005;**45**:988–995

[104] Mohammed AR, Tarpley L. Effects of night temperature, spikelet position and salicylic acid on yield and yield-related parameters of rice (*Oryza sativa* L.) plants. Journal of Agronomy and Crop Science. 2011;**197**:40–49

[105] Wang L, Chen S, Kong W, Li S, Archbold DD. Salicylic acid pretreatment alleviates chilling injury and affects the antioxidant system and heat shock proteins of peaches during cold storage. Postharvest Biology and Technology. 2006;**41**:244–251

[106] Shen H, Zhao B, Xu J, Zheng X, Huang W. Effects of salicylic acid and calcium chloride on heat tolerance in *Rhododendron* 'Fen Zhen Zhu'. Journal of the American Society for Horticultural Science 2016;**141**:363–372

[107] Saleh AA, Dina Z, Elish AA. Role of heat shock and salicylic acid in antioxidant homeostasis in mungbean (*Vigna radiata* L.) plant subjected to heat stress. American Journal of Plant Physiology. 2007; **2**:344–355

[108] Farooq M, Aziz T, Basra SMA, Cheema MA, Rehman H. Chilling tolerance in hybrid maize induced by seed priming with salicylic acid. Journal of Agronomy and Crop Science. 2008;**194**:161–168

[109] Larkindale J, Knight MR. Protection against heat stress-induced oxidative damage in *Arabidopsis* involves calcium, abscisic acid, ethylene, and salicylic acid. Plant Physiology. 2002;**128**:682–695

[110] Hasanuzzaman M, Fujita M. Heavy metals in the environment: current status, toxic effects on plants and possible phytoremediation. In: Anjum NA, Pereira MA, Ahmad I, Duarte AC, Umar S, Khan NA (eds) Phyto-technologies: remediation of environmental contaminants. Boca Raton: CRC Press; 2012. pp. 7–73

[111] Sharma P, Dubey RS. Involvement of oxidative stress and role of antioxidative defense system in growing rice exposed to toxic levels of aluminium. Plant Cell Reports. 2007;**26**:2027–2038

[112] Dubey RS. Metal toxicity, oxidative stress and antioxidative defense system in plants. In: Gupta SD (ed) Reactive oxygen species and antioxidants in higher plants. Boca Raton: CRC Press. 2011; pp. 177–203

[113] Fodor F. Physiological responses of vascular plants to heavy metals. In: Prasad MNV, Strzalka K (eds) Physiology and biochemistry of metal toxicity and tolerance in plants. Dordrecht: Kluwer. 2002; pp. 149–177

[114] Poschenrieder C, Barceló J. Water relations in heavy metal stressed plants. In: Prasad MNV(ed) Heavy metal stress in plants. Berlin: Springer; 2004. pp. 249–270

[115] Odjegba VJ. Exogenous salicylic acid alleviates arsenic toxicity in *Arabidopsis thaliana*. Indian Journal of Innovations and Developments. 2012;**1**:515–522

[116] Bai X, Dong Y, Kong J, Xu L, Liu S. Effects of application of salicylic acid alleviates cadmium toxicity in perennial ryegrass. Plant Growth Regulation. 2015;**75**:695–706

[117] Wael MS, Mostafa R, Taia AAE, Saad MH, Magdi TA. Alleviation of cadmium toxicity in common bean (*Phaseolus vulgaris* L.) plants by the exogenous application of salicylic acid. The Journal of Horticultural Science and Biotechnology. 2015;**90**:83–91

[118] Moussa HR, El-Gamal SM. Effect of salicylic acid pretreatment on cadmium toxicity in wheat. Biologia Plantarum. 2010;**54**:315–320

[119] Belkhadi A, Hediji H, Abbes Z, Nouairi I, Barhoumi Z, Zarrouk M, Chaïbi W, Djebali W. Effects of exogenous salicylic acid pre-treatment on cadmium toxicity and leaf lipid content in *Linum usitatissimum* L. Ecotoxicology and Environmental Safety. 2010;**73**:1004–1011

[120] Zhang F, Zhang H, Xia Y, Wang G, Xu L, Shen Z. Exogenous application of salicylic acid alleviates cadmium toxicity and reduces hydrogen peroxide accumulation in root apoplasts of *Phaseolus aureus* and *Vicia sativa*. Plant Cell Reports. 2011;**30**:1475–1483

[121] Agami RA, Mohamed GF. Exogenous treatment with indole-3-acetic acid and salicylic acid alleviates cadmium toxicity in wheat seedlings. Ecotoxicology and Environmental Safety. 2013;**94**:164–171

[122] Idrees M, Naeem M, Aftab T, Khan MMA, Moinuddin. Salicylic acid restrains nickel toxicity, improves antioxidant defence system and enhances the production of anticancer alkaloids in *Catharanthus roseus* (L). Journal of Hazardous Materials. 2013;**252**:367– 374

[123] Kazemi N, Khavari-Nejad RA, Fahimi H, Saadatmand S, Nejad-Sattari T. Effects of exogenous salicylic acid and nitric oxide on lipid peroxidation and antioxidant enzyme activities in leaves of *Brassica napus* L. under nickel stress. Scientia Horticulturae. 2010;**126**:402–407

[124] Mostofa MG, Fujita M. Salicylic acid alleviates copper toxicity in rice (*Oryza sativa* L.) seedlings by up-regulating antioxidative and glyoxalase systems. Ecotoxicology. 2013;**22**:959–973

[125] Lopez-Orenes A, Martínez-Perez A, Calderon AA, Ferrer MA. Pb-induced responses in *Zygophyllum fabago* plants are organ-dependent and modulated by salicylic acid. Plant Physiology and Biochemistry. 2014;**84**:57–66

[126] Song W, Zheng A, Shao H, Chu L, Brestic M, Zhang Z. The alleviative effect of salicylic acid on the physiological indices of the seedling leaves in six different wheat genotypes under lead stress. Peeters Online Journals. 2012;**5**:486–493

[127] Guo Q, Meng L, Mao P, Jia Y, Shi Y. Role of exogenous salicylic acid in alleviating cadmium induced toxicity in Kentucky bluegrass. Biochemical Systematics and Ecology. 2013;**50**:269–276

[128] Eraslan F, Inal A, Pilbeam DJ, Gunes A. Interactive effects of salicylic acid and silicon on oxidative damage and antioxidant activity in spinach (*Spinacia oleracea* L. cv. Matador) grown under boron toxicity and salinity. Plant Growth Regulation. 2008;**55**:207–219

[129] Eraslan F, Inal A, Gunes A, Alpaslan M. Impact of exogenous salicylic acid on the growth, antioxidant activity and physiology of carrot plants subjected to combined salinity and boron toxicity. Scientia Horticulturae. 2007;**113**:120–128

[130] Parashar A, Yusuf M, Fariduddin Q, Ahmad A. Salicylic acid enhances antioxidant system in *Brassica juncea* grown under different levels of manganese. International Journal of Biological Macromolecules 2014;**70**:551–558

[131] Shi Q, Zhu Z. Effects of exogenous salicylic acid on manganese toxicity, element contents and antioxidative system in cucumber. Environmental and Experimental Botany. 2008;**63**:317–326

[132] Lan T, You JF, Kong L, Yu M, Liu M, Yang Z. The interaction of salicylic acid and Ca^{2+} alleviates aluminum toxicity in soybean (*Glycine max* L.). Plant Physiology and Biochemistry. 2016;**98**:146–154

[133] Wang C, Zhang S, Wang P, Hou J, Qian J, Ao Y, Lu j, Li L. Salicylic acid involved in the regulation of nutrient elements uptake and oxidative stress in *Vallisneria natans* (Lour.) Hara under Pb stress. Chemosphere. 2011;**84**:136–142

[134] Zhou ZS, Guo K, Elbaz AA, Yang ZM. Salicylic acid alleviates mercury toxicity by preventing oxidative stress in roots of *Medicago sativa*. Environmental and Experimental Botany. 2009;**65**:27–34

[135] Sinha P, Shukla AK, Sharma YK. Amelioration of heavy-metal toxicity in cauliflower by application of salicylic acid. Communication in Soil Science and Plant Analysis. 2015;**46**:1309–1319

[136] Popova LP, Maslenkova LT, Yordanova RY, Ivanova AP, Krantev AP, Szalai G, Janda T. Exogenous treatment with salicylic acid attenuates cadmium toxicity in pea seedlings. Plant Physiology and Biochemistry. 2009;**47**:224–231

[137] Shi GR, Cai QS, Liu QQ, Wu L. Salicylic acid-mediated alleviation of cadmium toxicity in hemp plants in relation to cadmium uptake, photosynthesis, and antioxidant enzymes. Acta Physiologiae Plantarum. 2009;**31**:969–977

[138] Song W, Yang H, Shao H, Zheng A, Brestic M. The alleviative effects of salicylic acid on the activities of catalase and superoxide dismutase in malting barley (*Hordeum uhulgare* L.) seedling leaves stressed by heavy metals. Clean – Soil Air Water. 2014;**42**: 88–97

[139] Pinto DM, Blande JD, Souza SR, Nerg AM, Holopainen JK. Plant volatile organic compounds (VOCs) in ozone (O_3) polluted atmospheres: The ecological effects. Journal of Chemical Ecology 2010;**36**:22–34

[140] Meehl GA, Stocker TF, Collins WD, Friedlingstein P, Gaye AT, Gregory JM, Kitoh A, Knutti R, Murphy JM, Noda A, Raper SCB, Watterson IG, Weaver AJ, Zhao ZC. Global Climate Projections. In: Solomon S, Qin D, Manning M, Chen Z, Marquis M, Averyt KB, Tignor M, Miller HL (eds) Climate Change 2007: The Physical Science Basis. Contribution of Working Group I to the Fourth Assessment Report of the Intergovernmental Panel on Climate Change. Cambridge, New York: Cambridge University Press; 2007. pp. 749–845

[141] Nagatoshi Y, Mitsuda N, Hayashi M, Inoue S, Okuma E, Kubo A, Murata Y, Seo M, Saji H, Kinoshita T, Ohme-Takagi M. GOLDEN 2-LIKE transcription factors for chloroplast development affect ozone tolerance through the regulation of stomatal movement. Proceedings of the National Academy of Sciences. 2016;**113**:4218–4223

[142] Frohnmeyer H, Staiger D. Ultraviolet-B radiation-mediated responses in plants. Balancing damage and protection. Plant Physiology. 2003;**133**:1420−1428

[143] Müller-Xing R, Xing Q, Goodrich J. Footprints of the sun: Memory of UV and light stress in plants. Frontiers in Plant Science. 2014;**5**:474. doi: 10.3389/fpls.2014.00474.

[144] Janda M, Ruelland E. Magical mystery tour: Salicylic acid signaling. Environmental and Experimental Botany. 2015;**114**:117–128

[145] Wang J, Lv M, Islam F, Gill RA, Yang C, Ali B, Yan G, Zhou W. Salicylic acid mediates antioxidant defense system and ABA pathway related gene expression in *Oryza sativa* against quinclorac toxicity. Ecotoxicology and Environmental Safety. 2016;**133**:146–156

[146] Chen, YE, Cui, JM, Li GX, Yuan M, Zhang ZW, Yuan S, Zhang HY. Effect of salicylic acid on the antioxidant system and photosystem II in wheat seedlings. Biologia Plantarum. 2016;**60**:139–147

[147] Belkadhi A, Haro AD, Soengas P, Obregon S, Cartea ME, Chaibi W, Djebali W. Salicylic acid increases tolerance to oxidative stress induced by hydrogen peroxide accumulation in leaves of cadmium-exposed flax (*Linum usitatissimum* L.). Journal of Plant Interactions. 2014;**9**:647–654

[148] Liu W, Zhang Y, Yuan X, Xuan Y, Gao Y, Yan Y. Exogenous salicylic acid improves salinity tolerance of *Nitraria tangutorum*. Russian Journal of Plant Physiology. 2016;**63**:132–142

[149] Leslie CA, Romani RJ. Inhibition of ethylene biosynthesis by salicylic acid. Plant Physiology. 1988; **88**:833–837

[150] Hansen H, Grossmann K. Auxin-induced ethylene triggers abscisic acid biosynthesis and growth inhibition. Plant Physiology 2000;**124**:1437–1448

[151] Tewari RK, Paek KY. Salicylic acid-induced nitric oxide and ROS generation stimulate ginsenoside accumulation in *Panax ginseng* roots. Journal of Plant Growth Regulation. 2011;**30**:396–404

[152] Alavi SMN, Arvin MJ, Kalantari KM. Salicylic acid and nitric oxide alleviate osmotic stress in wheat (*Triticum aestivum* L.) seedlings. Journal of Plant Interactions. 2016;**9**:683–688

Chapter 4

Seed Dormancy: The Complex Process Regulated by Abscisic Acid, Gibberellins, and Other Phytohormones that Makes Seed Germination Work

Anna Skubacz and Agata Daszkowska-Golec

Additional information is available at the end of the chapter

http://dx.doi.org/10.5772/intechopen.68735

Abstract

Seed dormancy is one of the most important adaptive mechanisms in plants, which protects seeds from precocious germination in the presence of the inappropriate conditions for growth continuation. Numerous environmental and molecular signals regulate seed dormancy. Maintenance or release of seed dormancy is dependent on light, temperature, and water availability. Precise response of seeds to environmental factors is mediated by different phytohormonal pathways. ABA is considered as a main phytohormone regulating seed dormancy induction and maintenance. ABA- and GA-responsive components, ensure crosstalk between the GA and ABA pathways and enable seed response adequate to the environment. Phytohormonal regulation mechanism of seed dormancy is similar in dicot and monocot plants. Recently, it is suggested that other phytohormones, such as auxin, jasmonates, brassinosteroids, and ethylene, also take part in seed dormancy regulation. Auxin regulators, enhance ABA action and positively influence seed dormancy. However, jasmonates, brassinosteroids, and ethylene reduce seed dormancy level. Here, we describe recent advances in understanding the complex process of seed dormancy regulated by many phytohormonal pathways and their components. Seed dormancy studies can help obtain crop varieties producing seeds with the most desirable timing of germination.

Keywords: seed dormancy, germination, abscisic acid, gibberellic acid, phytohormone crosstalk

1. Introduction

Seed dormancy is defined as the inability of seeds to germinate under favorable conditions. The quiescent stage of seeds enables their survival during the adverse period for further seedling development. The high level of seed dormancy is considered as a negative trait due to germination retardation and reduction in the length of the growing season. On the other hand, low level of seed dormancy leads to preharvest sprouting (PHS) and yield loss. Thus, the varieties with medium value of seed dormancy are the most desirable [1–4]. Seed dormancy is considered as a quantitative trait under the control of the genetic and environmental signals. The primary dormancy is induced during seed maturation, and its expression occurs mainly in freshly harvested seeds in order to prevent precocious seed germination. After-ripening, which is dry seeds' storage at room temperature, can reduce primary seed dormancy [1]. The secondary dormancy can be induced in the presence of unfavorable conditions even in initially nondormant seeds [5–7]. Environmental conditions such as cold or heat temperature (stratification), light, nitrate (NO^{3-}), and nitric oxide (NO) can break the dormancy stage [1, 3, 6, 8, 9]. The level of seed dormancy depends on the season of a year. Deep dormancy is associated with sensing slow seasonal changes in winter. Shallow dormancy senses rapid condition changes in summer [10].

Induction and release of seed dormancy is mainly under the control of abscisic acid (ABA) and gibberellic acid (GA). ABA promotes seed dormancy and germination inhibition. Action of ABA is counteracted by GA, which promotes seed germination at appropriate time. The balance between ABA and GA is regulated by environmental conditions (light, temperature) and endogenous signals [4, 6, 7, 11]. Other phytohormones, such as auxin, brassinosteroids, and ethylene, modulate the interaction between ABA and GA in the regulation of seed dormancy [2, 4, 12].

Seed dormancy in cereals is established during seed development; however, the time of seed dormancy release can be different. Some varieties loose dormancy when the harvest maturity is reached. There are also varieties ready for germination after seed physiological maturity (fully developed, but not dried seeds). In cereals, such as barley (*Hordeum vulgare*), wheat (*Triticum aestivum*), and sorghum (*Sorghum bicolor*), the switch between physiological and harvest maturity is related to ABA decrease [7].

Here, we discuss the genetic and molecular bases of seed dormancy entrance and breaking in Arabidopsis and monocot plants, considering the action of components belonging to ABA, GA, and other phytohormone pathways. Additionally, the influence of environmental cues on ABA- and GA-related genes is described.

2. Role of ABA metabolism and signaling in maintaining seed dormancy

ABA is considered as a crucial phytohormone for seed dormancy establishment and maintenance. Many of the ABA metabolism- and signaling-related genes play a crucial role in the control of seed dormancy.

2.1. ABA biosynthesis and catabolism activity in the regulation of seed dormancy

ABA produced in the embryo is fundamental for the promotion of seed dormancy. ABA synthesized in maternal tissues or ABA applied externally is not able to induce seed dormancy [13]. However, Kanno et al. [14] showed that ABA produced by maternal tissues can be transported to the embryo in order to take part in seed dormancy induction. ABA biosynthesis is catalyzed in several steps, and the rate-limiting reaction is mediated by carotenoid cleavage dioxygenase (NCED) [15, 16].

Many ABA biosynthesis genes are implicated in the regulation of seed dormancy in Arabidopsis. *NCED6* and *NCED9* are considered as the key ABA biosynthesis genes for induction of seed dormancy. They are expressed specifically during seed development. Double mutant *nced6*/*nced9* shows reduced seed dormancy [17]. Additionally, overexpression of *NCED6* results in an increase in the ABA content in seeds and in the inhibition of precocious germination [18]. *NCED5* is also described as a seed dormancy regulator (**Figure 1**) [19]. Other enzymes necessary for ABA biosynthesis and seed dormancy are encoded by *ABA deficient 2* (*ABA2*) and *abscisic aldehyde oxidase 3* (*AAO3*). *aba2-2* and *aao3* mutants show a reduced ABA content and similar disorders in seed dormancy as *nced6*/*nced9* (**Figure 1**) [14, 20]. ABA level in seeds depends also on degradation process. The catabolism of ABA is mediated by ABA8'hydroxylase encoded by *cytochrome P450* (*CYP707A*) genes [15, 16]. The activity of *CYP707A* genes is related to the loss of seed dormancy. *CYP707a* mutants show higher level of seed dormancy than the wild type (WT), especially the *CYP707A2*. The expression of *CYP707A2* is induced in seeds during imbibition. Furthermore, *CYP707A2* activity and after-ripening show a positive relationship. Therefore, *CYP707A2* is proposed to be responsible mainly for ABA degradation during release of seed dormancy and the germination process (**Figure 1**) [21, 22]. The other *CYP707A* genes, *CYP707A1* and *CYP707A3*, also take part in ABA catabolism in seeds; however, their role in breaking dormancy is minor [22, 23].

The regulation of ABA metabolism genes plays also a very important role in seed dormancy of monocot plants. In rice (*Oryza sativa*), the expression of *OsNCED2* is activated at the early or the late stage of seed development, in dormant and nondormant cultivars, respectively. The different times of ABA biosynthesis in seeds could result in a high or low dormancy level [24]. In barley, the expression pattern of *HvNCED* genes in developing grains shows the higher level of *HvNCED2* transcript in comparison to *HvNCED1* [25–27]. Moreover, *HvNCED2* activation in the field is independent of weather conditions, in contrary to *HvNCED1* and *HvABA8'OH1*/*HvCYP707A1*. On the other hand, the induction of *HvABA8'OH1* expression occurs in after-ripened seeds, but not in the dormant seeds during imbibition. Thus *HvNCED2* seems to play a more significant role in ABA biosynthesis and in the preventing of preharvest sprouting than *HvNCED1*. Furthermore, *HvABA8'OH1* activity mediates dormancy breaking [25].

Barley seed dormancy is associated with the presence of glumellae (lemma and palea). It was shown that dehulled grains have no induction of *HvNCED1*, *HvNCED2*, and *HvABA8'OH1* genes. The contrary reaction was observed in whole, dormant grains [28]. The induction of secondary dormancy in barley is also dependent on ABA metabolism genes. While *HvNCED1*

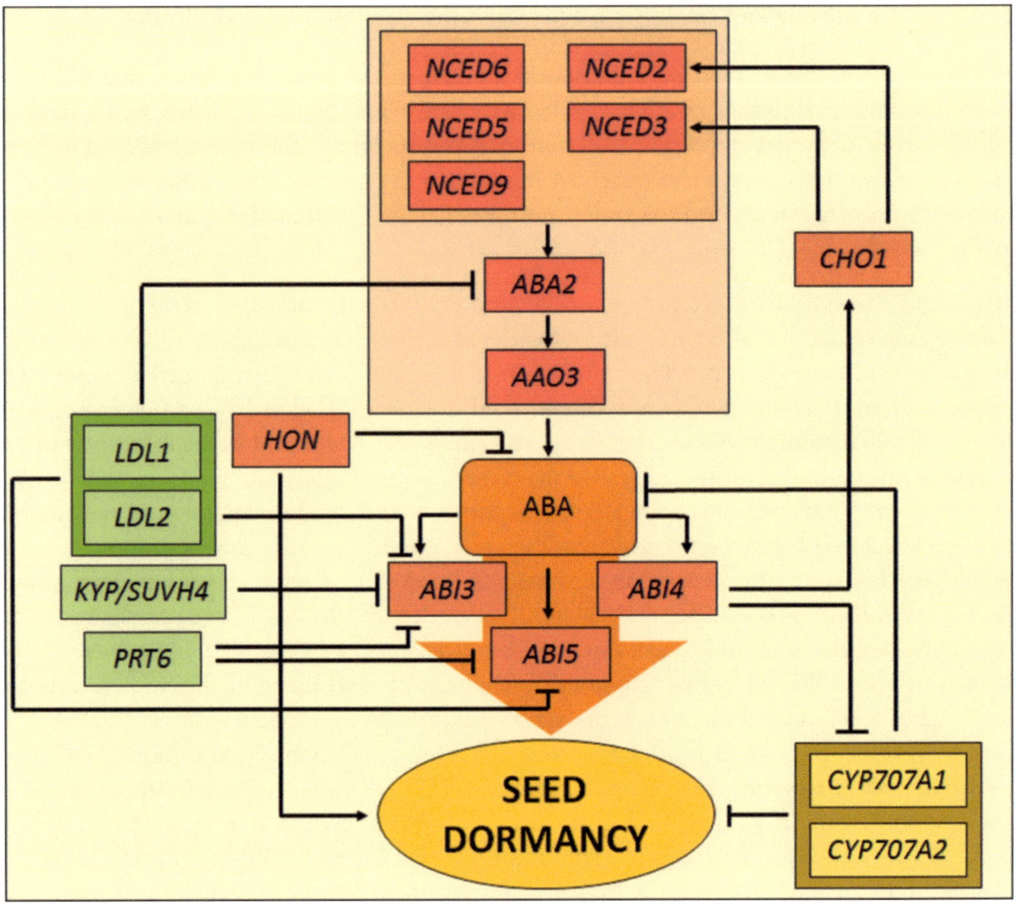

Figure 1. Probable function of ABA-related genes in seed dormancy promotion. Seed dormancy is positively regulated by ABA biosynthesis genes: *carotenoid cleavage dioxygenase 2 (NCED2), NCED3, NCED5, NCED6, NCED9, ABA2 ABA deficient* 2 (*ABA2*), and *abscisic aldehyde oxidase* 3 (*AAO3*). Genes encoding ABA-related transcription factors, *ABA insensitive 3* (*ABI3*), *ABI4*, and *ABI5*, also promote seed dormancy. *HONSU* (*HON*) is a positive regulator of seed dormancy, but it represses ABA signaling. ABA catabolism genes, *cytochrome P450* (*CYP707A1*) and *CYP707A2*, are responsible for seed dormancy release. Other regulators like, *lysinespecific demethylase 1* (*LDL1*), *LDL2*, *kryptonite/SU(VAR)3-9 homolog 4* (*KYP/ SUVH4*), *proteolysis6 (PRT6)* negatively regulates the ABA pathway and seed dormancy. ABI4 modulates the expression of ABA biosynthesis genes (*NCED2*, *NCED3*) probably via *CHOTTO1* (*CHO1*) and ABA catabolism genes (*CYP707A1*, *CYP707A2*).

shows higher activation at 30°C, the transfer of grains to 20°C is associated with *HvNCED2* and *HvABA8′OH1* induction. Probably, the expression of secondary dormancy depends mostly on *HvNCED2*, whereas the promotion of *HvABA8′OH1* may be a response to immediate increase in the ABA level in seeds [26]. *HvNCED2* is also mostly expressed during hypoxia-related seed dormancy [29]. The after-ripening process is associated with the increased expression of *HvABA8′OH1* in coleorhizae [30–32]. The barley lines with silenced *HvABA8′OH1* expression show the increased ABA accumulation and seed dormancy level [31]. In *Brachypodium distachyon*, the higher expression of *BdNCED1* was observed in dormant grains in comparison to after-ripened grains. Contrarily, after ripening promoted the induction of *BdABA8′OH-1* at the second day of imbibition. Probably, *BdABA8′OH-1* plays a prominent role in the after-ripening process [33].

2.2. Regulation of seed dormancy via ABA signaling components

The core ABA signaling is mediated by pyrabactin resistance proteins/PYR-like proteins/regulatory components of ABA receptor (PYR/PYL/RCAR), phosphatase 2C (PP2C), SNF1-related protein kinase 2 (SnRK2), and abscisic acid responsive elements-binding factor (AREB) basic leucine zipper (bZIP) transcription factors [34–36]. In Arabidopsis, ABA signaling genes are also implicated in seed dormancy regulation. *ABA insensitive 1* (*ABI1*) encodes PP2C phosphatase, which acts as the negative regulator of ABA signaling [37]. *abi1* was described as the mutant with decreased seed dormancy level and better germination in the presence of ABA [38]. The other PP2C phosphatase, HONSU (HON), also represses ABA signaling, specifically in seeds. However, its role in seed dormancy is inconclusive. *HON* expression is associated with both, dormancy establishment and release (**Figure 1**) [39]. *ABI* genes, such as *ABI3*, *ABI4*, and *ABI5* encode crucial ABA-dependent transcription factors expressed in seeds. Expression of *ABI3*, *ABI4*, and *ABI5* is higher in dormant seeds than in seeds with reduced seed dormancy level (**Figure 1**) [40–42]. Among *ABI* genes, *ABI3* is the most substantial for seed dormancy establishment. *ABI3* is expressed in developing seeds. It regulates the accumulation of chlorophyll, anthocyanins, and storage proteins together with two other seed-related regulators, *FUSCA3* (*FUS3*) and *leafy cotyledon 1* (*LEC1*) [43, 44]. *abi3* mutant shows no seed dormancy, and immature seeds are able to germinate [45]. ABI3 is under direct regulation of WRKY DNA-binding protein 41 (WRKY41) during the establishment of primary seed dormancy. WRKY41 binds directly to *ABI3* promoter and induces its expression [46].

ABI4 is another ABA-activated transcription factor with APETALA 2 (AP2) domain, expressed in seeds. It takes part in the regulation of abiotic stress responses and different aspects of plant development [47]. *abi4* mutant germinates faster than the wild type without stratification. The expression analysis showed decreased activation of *NCED2* and *NCED3* in *abi4* seeds. Moreover, ABI4 binds to *CYP707A1* and *CYP707A2* promoters and represses their expression. It indicates the important role of ABI4 in seed dormancy maintenance (**Figure 1**) [41]. It is worth noting that *ABI4*, *NCED2*, and *NCED6* are under positive regulation of a common ABA-dependent regulator, myeloblastosis 96 (MYB96). The activation of *NCED2* and *NCED6* ensures ABA biosynthesis and seed dormancy promotion, whereas *ABI4* induction inhibits lipid breakdown and further seed germination [48]. One of the downstream target of ABI4 is *CHO1* (*CHOTTO 1*), encoding a transcription factor with double AP2 domain. *CHO1* acts also as a positive regulator of primary seed dormancy (**Figure 1**) [49, 50]. ABI5 is a bZIP transcription factor regulating ABA signaling in seeds [42]. The role of *ABI5* in seed dormancy regulation is not clear. *abi5* mutant shows a normal dormancy level [51]. However, many studies described below showed a distinct relationship between *ABI5* and seed dormancy [40, 52–54].

In monocot plants, the activation of ABA signaling is also associated with seed dormancy. The maize (*Zea mays*) ortholog of *ABI3*, the *viviparous 1* (*VP1*), is a crucial regulator of seed dormancy. *vp1* mutant shows premature embryo germination (vivipary) and reduced ABA sensitivity [55]. The overexpression of maize *VP1* in wheat induces increased seed dormancy and prevents preharvest sprouting [56]. Some rice varieties produce truncated versions of *OsVP1* transcript. There is a relation between incorrect transcripts' amount and preharvest sprouting. This phenomenon is associated with developmental stage: immature embryos accumulate a higher number of truncated transcripts than mature embryos [57]. Another gene, the *seed dormancy 4* (*SDR4*) is a rice quantitative trait locus (QTL) responsible for seed dormancy promotion in ABA-dependent manner. The *japonica* varieties have reduced dormancy and possess only SDR4-n allele, whereas more dormant varieties of *indica* type include SDR4-n and SDR4-k alleles. OsVP1 was shown to positively promote *SDR4* expression [58]. Other ABA-related genes also take part in seed dormancy maintenance. Expression analysis of sorghum grains with various dormancy level identified a set of differentially regulated ABA signaling genes. A dormant inbred line of sorghum showed increased expression of *SbABA-responsive protein kinase* (*SbPKABA1.1*), *SbABI1*, *SbVP1*, *SbABI4*, and *SbABI5* during grain imbibition. However, no induction of these genes in a nondormant inbred line was observed [59]. In barley, dormancy expression is associated with increased induction of *HvPKABA*, *HvVP1*, and *HvABI5* [28]. Probably, the general ABA-related mechanism of seed dormancy induction is similar in dicots and monocots.

2.3. Environmental cues and epigenetic modifications in the regulation of the ABA pathway

The expression of Arabidopsis ABA metabolism and signaling genes is regulated through environmental factors. The red (R) light pulse irradiation applied to the far-red (FR) light pulse pretreated, dark-imbibed seeds inhibits and induces the expression of *NCED6* and *CYP707A2*, respectively. It suggests that the ABA metabolism genes are under the control of PHYB (phytochrome B), which regulates germination in response to FR and R pulse light [60]. On the other hand, the blue light has a negative impact on the germination of dormant grains in cereals. The blue light–associated secondary dormancy induces *HvNCED1* and *HvNCED2* and weakly reduces *HvABA8'OH-1* expression in grains [61]. The activation of *HvNCED1* is under the regulation of phytochrome photoreceptor, cryptochrome 1 (HvCRY1). It indicates that ABA biosynthesis and catabolism take part in blue light–dependent regulation of seed dormancy [62, 63]. The temperature and NO also exert an impact on ABA pathway in Arabidopsis seeds. The high temperature promotes the expression of ABA biosynthesis genes in imbibed seeds, whereas NO positively regulates ABA signaling during seed dormancy breaking [52, 64]. NO action may be associated with N-end rule pathway, leading to degradation of proteins with destabilizing amino acid residues. NO and oxygen are sensed by N-end rule pathway with the participation of many protein regulators [65]. The components of N-end rule pathway, proteolysis 6 (PRT6) and arginyl-tRNA:protein arginyltransferase (ATE), regulate after-ripening, inhibit ABA signaling, and finally promote seed germination. PRT6 is E3 ligase promoting protein degradation via 26S proteasome. Some PRT6 substrates belong to the ABA pathway. As a result, ABA signaling is inhibited, and the activation of ABI3 and ABI5 is detained (**Figure 1**) [52].

The ABA metabolism and signaling genes are also regulated at epigenetic level during the establishment of seed dormancy. Kryptonite/SU(VAR) 3-9 homolog 4 (KYP/SUVH4) is responsible for histone H3 lysine 9 dimethylation. Repression of *ABI3* by KYP/SUVH4 is required to release seed dormancy (**Figure 1**) [66]. Moreover, the expression of *ABA2, ABI3,* and *ABI5* is downregulated through the action of two histone demethylases, lysinespecific demethylases 1 and 2 (LDL1 and LDL2). Thus, the activity of LDL1 and LDL2 ensures decrease in primary seed dormancy via negative regulation of ABA response (**Figure 1**) [67].

3. Gibberellins-mediated control of seed dormancy release and germination

A high level of gibberellins (GA) is needed for the counteraction of ABA activity in seeds. GA promotes seed dormancy release and radical protrusion during seed germination. The activation of GA-responsive genes induces cell wall–remodeling enzymes, such as endo-β-mannase, xyloglucan endotransglycolase, expansin, and β-1,3-mannase. Their activity leads to the weakening of the embryo-surrounding layers. Additionally, GA ensures the high-growth potential of the embryo [68].

3.1. Role of GA metabolism in seed dormancy break

GA biosynthesis takes place mainly in the radicle of the embryo, which in turn ensures germination progression [69]. Arabidopsis seed germination is associated with the regulation of GA metabolism genes. The highest expression of GA-biosynthesis genes, *gibberellin 3-oxidase 1* (*GA3ox1*), *GA20ox3*, and *ENT-kaurene oxidase 1* (*KO1*), was shown during the first 8 hours of imbibition [68]. The crucial role of GA in the breaking of seed dormancy was presented using a *ga requiring 1* (*ga1*) mutant in GA biosynthesis gene, *CPP synthase* (*CPS*). Interestingly, *ga1* capacity to germinate was renewed after removing testa and endosperm, without exogenous GA application. It was concluded that dormancy release and germination promotion was dependent on GA-ABA balance in the embryo and the embryo-surrounding layers of the seed [70]. The environmental factors, such as light and temperature, interact with GA biosynthesis and signaling, which in turn promotes seed germination. The expression of *GA3ox1* is activated by red light and cold. Additionally, the low temperature determines the *GA3ox1* expression localization in the embryonic axis and the aleurone layer [71, 72]. Contrarily, low temperature represses the expression of GA catabolism gene, *GA2ox2* [71]. Two bHLH (basic helix-loop-helix) transcription factors, spatula (SPT) and phytochrome interacting factor 3-like 5 (PIL5), regulate seed germination after cold stratification, including GA biosynthesis pathway. SPT represses germination before stratification, whereas PIL5 also acts as an inhibitor of germination, but after cold stratification in darkness. Both, SPT and PIL5, act through negative regulation of *GA3ox1* and *GA3ox2* [73]. Another transcription factor, DOF affecting germination1 (DAG1) was found to mediate PIL5 negative regulation of *GA3ox1*. PIL5 promotes the expression of *DAG1* in darkness. Furthermore, DAG1 protein binds directly to *GA3ox1* promoter, inhibits its expression, and blocks germination [74]. Contrary to cold, high temperature represses the expression of *GA20ox1*, *GA20ox2*, *GA20ox3*, *GA3ox1*, and *GA3ox2* during seed imbibition and blocks germination [64]. Similar to ABA-related genes, the expression of GA metabolism genes is regulated seasonally. *GA3ox2* activation is associated with summer, whereas *GA2ox2* is expressed in winter [10].

In barley and wheat, the expression of GA biosynthesis genes occurs during imbibition of nondormant seeds [31, 72]. The rapid increase in *HvGA3ox2* involved in GA biosynthesis was observed in the after-ripened grains during imbibition. The high expression level of *HvGA3ox2* is associated with *HvGA2ox2* activation [31]. The hypoxia-related secondary dormancy in barley is associated with the modulation of the GA pathway. Low oxygen concentration causes induction of *HvGA2ox3* and repression of *HvGA3ox1* and *HvGA20ox1* in dormant grains. The activity of GA-responsive gene, *HvEXPANSIN11* (*HvEXPA11*), is also repressed [29]. Similar reaction was observed during seed dormancy imposed by blue light. The negative regulation of the GA pathway occurred through the promotion of *HvGA2ox3* and *HvGA2ox5* and the repression of *HvGA3ox2* [61]. The relationship between the expression of GA metabolism genes and the induction of secondary dormancy at 30°C was also shown; however, the particular expression pattern depended on the embryo water content in barley. The embryo with high-water content (1.60–1.87 g H_2O g^{-1} DW) shows the higher expression of GA catabolism and signaling genes than the embryo with lower water content (0.45 g H_2O g^{-1} DW) [75].

GA metabolism genes are involved in seed dormancy regulation in other monocot species. In wheat, after-ripening causes induction of *GA20ox1* and *GA3ox2* [72]. The regulation of expression of GA synthesis and catabolism genes is more complex in rice. *OsGA20ox1*, *OsGA2ox5*, and *OsGA2ox6* expression pattern showed higher variability in a nondormant than in a dormant variety during seed development. Furthermore, the dormant variety accumulated less-active GA in seeds in comparison to the nondormant variety. It resulted in appropriate dormancy phenotype of analyzed cultivars [24]. Similar analysis was conducted in terms of immature grains of sorghum inbred lines with contrasting dormancy level. Higher expression of *SbGA20ox1* and *SbGA20ox3* was observed for a less-dormant line, whereas a strong induction of *SbGA2ox1* and *SbGA2ox3* was found in the line with higher dormancy [76]. To summarize, a proper regulation of GA biosynthesis and catabolism genes ensures the regulation of seed dormancy dependent on environmental conditions, both in dicot and monocot plants.

3.2. Action of GA signaling components in seed dormancy regulation

In Arabidopsis, GA signaling is mediated by GA insensitive dwarf1 (GID1) receptor. Overexpression of *GID1* promotes the release of seed dormancy. The impact of cold stratification and after ripening on *GID1* expression showed that imbibition at 4°C promoted expression of three *GID1* transcript forms: *GID1a*, *GID1b*, and *GID1c*, while after-ripening storage induced only *GID1b*. Thus, both mechanisms of seed dormancy loss seem to be regulated differently [77]. In sorghum, exogenous GA represses *SbGID1* in immature grains. It suggests the role of SbGID1 in negative feedback regulation of the GA pathway [76]. The sleepy1 (SLY1) is a F-box protein which enables 26S proteasome-mediated degradation of

DELLA proteins in the presence of active GA [78]. DELLA proteins act as repressors of GA signaling. *sly1* mutant shows reduced germination, even after the application of exogenous GA. It indicates that SLY1 is the crucial regulator of seed germination [79]. Another mutant related to GA signaling, *cts* (*comatose*), maintains seed dormancy even after stratification or after ripening. CTS functions as a peroxisomal ABC transporter and seems to be crucial for seed dormancy release [80]. The proper regulation of DELLA proteins is crucial for seed germination. Simultaneous deactivation of *repressor of GA* (*RGA*), *RGA-like 1* (*RGL1*), *RGL2*, and *gibberellic acid insensitive* (*GAI*) results in insensitivity to GA and light during germination. It indicates that DELLA proteins integrate environmental cues into GA signaling [81]. Among them, RGL2 seems to play a more important role in seed germination than other DELLAs. Thermoinhibition of seed germination demands activity of *RGL2*, which suggests its crucial role in the regulation of GA signaling in seeds [64]. Moreover, *GID1* transcripts are under control of RGL2 during cold stratification and after ripening. The RGL2 can promote or inhibit *GID1* expression according to a particular *GID1* transcript form and surrounding conditions during dormancy loss [77]. *RGL2* activity associates with the regulation of shallow dormancy. Its expression is promoted during summer time [10]. Another negative regulator of GA signaling is *spindly* (*SPY*). The *spy* mutant demands the lower amount of GA to break seed dormancy and continue germination. *SPY* encodes *O*-linked *N*-acetylglucosamine (*O*-GlcNAc) transferase which probably glycosylates components of GA signaling. SPY acts upstream of RGA through the modulation of its activity through *O*-GlcNAc modification [82].

3.3. The role of essential seed dormancy regulator, DOG1, in GA pathway regulation

Delay of germination 1 (*DOG1*) is considered as the crucial, positive regulator of seed dormancy with unknown function. Expression of *DOG1* is seed specific, and *dog1* mutant shows disturbed seed dormancy in Arabidopsis [83]. Similar to ABA-related genes, *DOG1* is under negative epigenetic regulation mediated by KYP/SUVH4, LDL1, and LDL2, which as a result reduces primary dormancy [66, 67]. *DOG1* expression is related to deep dormancy during winter season [10]. Recently, the role of DOG1 in temperature-dependent coat dormancy through GA metabolism regulation was shown. DOG1 differently regulates the expression of GA biosynthesis genes, such as *GA3ox1* and *GA20ox*, at 18 and 24°C. This leads to the inhibition of genes encoding cell wall remodeling enzymes: *expansin 2* (*EXPA2*), *EXPA9*, *xyloglucan endo-transglycosylase 19* (*XTH19*) but only at 24°C. Therefore, DOG1 regulates the appropriate time of germination according to environment temperature [84].

4. ABA and GA crosstalk during seed dormancy

The seed dormancy maintenance or release and further promotion of the seed germination process are regulated by ABA and GA balance [1, 2, 12]. ABA-mediated repression of GA biosynthesis enables the positive regulation of seed dormancy [60]. Many molecular interactions between ABA and GA pathways enable precise regulation of seed response according to environmental conditions.

4.1. Activity of ABA and GA metabolism genes ensures the ABA-GA interaction

There is the relationship between ABA and GA biosynthesis in Arabidopsis. ABA-deficient mutant, *aba2-2*, shows the higher expression of *GA3ox1* and *GA3ox2* than the wild type [60, 64]. Interestingly, *AAO3* and *ABA2* expression were detected in a radicle, whereas *GA3ox2* in hypocotyl. It suggests that the places of ABA and GA biosynthesis are different in seeds [60]. *aba2-2* shows also the reduced expression of *SPY* during seed imbibition in the presence of high temperature. Therefore, the negative regulator of GA signaling, *SPY*, is under the positive action of ABA (**Figure 2**) [64]. NCED9 negatively influences GA biosynthesis. The application of paclobutrazol, GA biosynthesis inhibitor, causes better germination of *nced9* than the wild type. It is an evidence that ABA biosynthesis modulates GA pathway in seeds [85].

4.2. ABA-GA crosstalk depends on ABI transcription factors and DELLA proteins in seeds

ABA and GA signaling components are involved in the ABA-GA crosstalk in Arabidopsis seeds. ABI4 exerts action on GA biosynthesis genes. In *abi4* mutant, the expression of *GA3*,

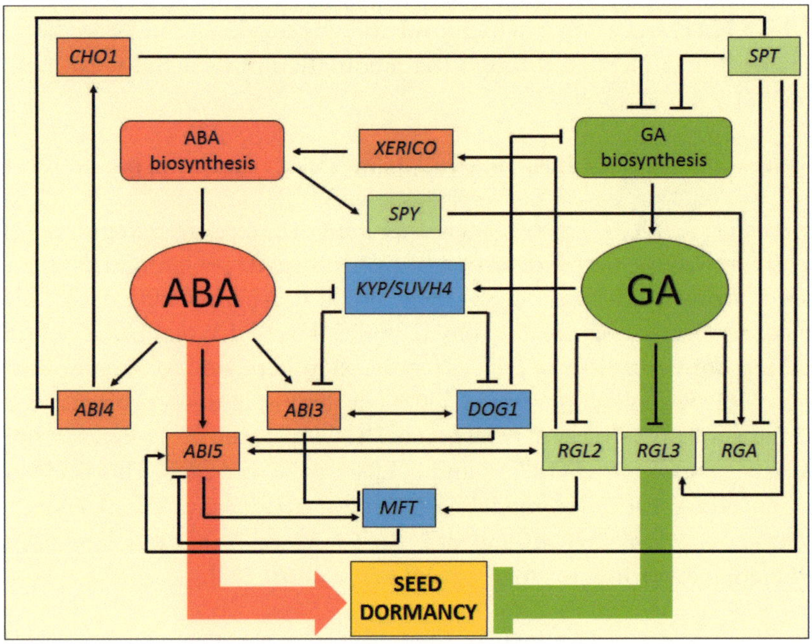

Figure 2. Model for seed dormancy regulation by ABA-GA crosstalk. ABA-mediated promotion of seed dormancy and GA-related release of seed dormancy are possible through ABA-GA interactions. The seed dormancy regulator, *mother of FT and TFL 1 (MFT1)*, is promoted by ABA insensitive 5 (ABI5) and RGA-like 2 (RGL2), but ABI3 downregulates its expression. ABI5 and RGL2 positively regulate reciprocal expression. RGL2 also promotes *XERICO* and ABA biosynthesis. Repressor of GA biosynthesis, delay of germination 1 (DOG1), activates *ABI3* and *ABI5*. GA biosynthesis is inhibited by spatula (SPT) and ABI4 via chotto1 (CHO1) activity. SPT also represses the expression of *ABI4* and *RGA repressor of GA (RGA)* but promotes *ABI5* and *RGL3*. *Spindly* (*SPY*), a negative regulator of GA signaling, is promoted by ABA. Modulation of ABA and GA responses also includes an epigenetic regulator, kryptonite/SU(VAR)3-9 homolog 4 (KYP/SUVH4).

GA3ox1, *GA20ox1*, *GA20ox2*, *GA20ox3*, *ENT-kaurenoic acid oxydase 1* (*KAO1*), and *KAO2* genes is upregulated in imbibed seeds. The *abi4* seeds also accumulate more GA [41]. *CHO1* acts downstream of *ABI4* in seed dormancy regulation, and its activity leads to the repression of GA biosynthesis genes (**Figure 2**) [49]. RGL2 seems to be one of the most important GA-related component acting in ABA-GA crosstalk in seeds. The positive interaction between RGL2 and ABA biosynthesis through *XERICO* was described (**Figure 2**) [86]. Moreover, *RGL2* and *ABI5* positively regulate reciprocal expression during seed germination (**Figure 2**) [87]. Recently, the cooperation of NF-YC transcription factor with RGL2 was identified during the regulation of *ABI5* expression in seeds [88].

Coat-mediated dormancy is also related to RGL2 action. RGL2 promotes ABA biosynthesis in endosperm, then coat-derived ABA is released to the embryo, where it ensures the expression of *ABI5* and in consequence germination inhibition [40]. ABA and GA signaling genes are under the control of the negative regulator of GA biosynthesis, SPATULA (SPT): *ABI5* and *RGL3* are promoted, whereas *ABI4* and *RGA* are repressed by SPT. It suggests the universal role of SPT in seed dormancy induction and release through complex influence on ABA and GA pathways (**Figure 2**) [89]. Induction of secondary dormancy through seed imbibition in darkness at 25°C is associated with changes in GA content and signaling. However, this process also includes positive action of RGL2 on *ABI5*. It suggests that ABA-GA crosstalk is also important for entrance into secondary dormancy [90]. Epigenetic modifications are implicated in ABA-GA interaction. *KYP/SUVH4* is promoted by GA and repressed by ABA. Regarding the role of KYP/SUVH4 in the regulation of *ABI3* and *DOG1* expression, this histone methylotransferase is also a part of ABA-GA interaction (**Figure 2**) [66].

The interaction between ABI transcription factors and GA catabolism genes was described in monocot plants. In sorghum, SbABI4 and SbABI5 are able to bind with coupling element 1 (CE1) and ABA responsive element (ABRE), respectively, that are present in *SbGA2ox3* promoter and subsequently promote its expression. ABA-dependent activation of GA catabolism can promote seed dormancy in grains [53].

4.3. Seed dormancy regulators, MFT and DOG1, are a part of the ABA-GA crosstalk

Mother of FT and TFL 1 (MFT) is one of the crucial regulators of seed dormancy enabling the interaction between ABA and GA signaling in Arabidopsis. MFT negatively regulates ABA signaling and seed dormancy, which in turn leads to germination. Its expression is repressed by ABI3 but promoted by RGL2. MFT also ensures a negative feedback loop in ABA signaling through the repression of *ABI5* transcription, whereas ABI5 induces *MFT* expression (**Figure 2**) [91]. However, the role of MFT is not completely clear. The wheat ortholog of MFT, TaMFT, acts in an opposite way in seed dormancy regulation. The increased expression of *TaMFT* is related to the lower germination index, and *TaMFT* overexpression causes inhibition of precocious germination of isolated embryos. Low temperature during seed development is associated with a higher level of dormancy. Under such environmental conditions, the activation of *TaMFT* was observed during seed development [92]. Probably, the precise role of MFT in seed dormancy is different in dicots and monocots.

The role of DOG1, the GA-related regulator of seed dormancy, was also described in ABA signaling in seeds. *ABI5* is positively promoted by DOG1, which in turn leads to the regulation of many *late embryogenesis abundant* (*LEA*) and *heat shock protein* (*HSP*) genes. Moreover, the double-mutant *abi3-1/dog1-1* shows the lower sensitivity to ABA than *abi3-1*, and in control condition, it produces mature dry green seeds. It suggests the positive relationship between *DOG1* and *ABI3; therefore,* DOG1 may be responsible for ABA-GA interactions in seeds (**Figure 2**) [54].

5. The emerging role of auxin, jasmonates, brassinosteroids, and ethylene in seed dormancy regulation

ABA and GA are not the only phytohormonal regulators of seed dormancy establishment and release. Their action is modulated by other phytohormones, such as auxin, jasmonates (JA), brassinosteroids (BR), and ethylene.

5.1. Action of auxin pathway components in seeds

Auxin promotes seed dormancy release and germination. Constitutive induction of auxin biosynthesis in *iaaM-OX* line inhibits precocious germination in Arabidopsis. Contrarily, the switched off activity of *auxin response factor 10* (*ARF10*) and *ARF16*, auxin-dependent transcription factors, in *arf10/arf16* double mutant, causes faster precocious germination than in the wild type. The role of auxin in the control of seed dormancy includes the action of *ABI3*. The double mutants, *abi3-1/iaaM-OX* and *abi3-1/99999mARF16* (line resistant to miR160), show the reduced dormancy phenotype. Therefore, the activation of auxin signaling promotes *ARF10* and *ARF16*, which in turn induces *ABI3* and seed dormancy (**Table 1**) [93]. Analysis of after-ripened wheat grains showed increased expression of *TaIAA-alanine resistant 3* (*TaIAR3*) encoding hydrolase releasing IAA from conjugates. It was observed in parallel with the higher IAA level in seeds during imbibition. Probably, seed dormancy release may be associated with the increased auxin content in seeds of monocot plants. Furthermore, *TaAuxin-resistant 1* (*TaAXR1*), *TaUbiquitin-related protein 1* (*TaRUB1*), and *TaARF2* were also upregulated in after-ripened wheat grains. TaAXR1 is associated with AUX/IAA proteasome-mediated degradation, whereas TaRUB1 is related to ubiquitin action. The higher expression of *TaAXR1* and *TaRUB1* can exert a negative impact on auxin signaling (**Table 1**) [72].

5.2. Dual role of jasmonic acid in seed dormancy regulation

The role of JA (Jasmonic Acid) in seed dormancy is ambiguous. The increased JA content was detected in nondormant Arabidopsis seeds. Probably, the decrease of JA content during imbibition in nondormant seeds is associated with germination promotion [94]. Application of JA precursor, 12-oxo-phytodienoic acid (OPDA) promotes the expression of *ABA1, ABI5,* and *RGL2* in after-ripened seeds and inhibits seed germination. OPDA also exerts a regulatory action on the crucial seed dormancy component, *MFT* [95]. The opposite effect of JA on seed

Phytohormonal pathway	Regulator	Function	Role in seed dormancy regulation	References
Auxin	ARF10 ARF16	Auxin-related transcription factors	Promotion of *ABI3* expression and seed dormancy	[93]
	TaIAR3 TaARF2 TaAXR1 TaRUB1	Releasing auxin from conjugates Auxin-related transcription factor Aux/IAA proteasome degradation-associated protein Ubiquitin pathway-associated protein	Seed dormancy release	[93]
Jasmonic Acid	TaAOS TaKAT3 TaLOX5	JA biosynthesis	Seed dormancy release	[93]
	TaAOC TaAOS	JA biosynthesis	Seed dormancy release via repression of *TaNCED1* and *TaNCED2*	[96] [97]
Brassinosteroids	TaBIN2	Negative regulator of BR signaling with kinase activity	Seed dormancy promotion via ABI5 activation	[99, 100]
	TaDET2 TaDWF4 TaBSK2	BR biosynthesis Positive regulator of BR signaling with kinase activity	Seed dormancy release	[100]
Ethylene	ACO ETR1 EIN2	Ethylene biosynthesis Ethylene receptors	Seed dormancy release Seed dormancy release through the regulation of ABA metabolism genes	[101] [104]

Note: auxin response factor (ARF), IAA-alanine resistant 3 (IAR3), auxin-resistabt 1 (AXR1), ubiquitin-related protein 1 (RUB1), allene oxide synthase (AOS), 3-ketoacyl coenzyme A (KAT3), lipoxygenase 5 (LOX5), allene oxide cyclase (AOC), brassinosteroid insensitive 2 (BIN2), de-etiolated 2 (DET2), DWARF 4 (DWF4), br signaling kinase 2 (BSK2), 1-aminocyclopropane-1-carboxylic acid oxidase (ACO), ethylene triple response 1 (ETR1), ethylene insensitive 2 (EIN2).

Table 1. Regulators of auxin, jasmonic acid, brassinosteroid, and ethylene pathways in seed dormancy promotion or release.

dormancy exists in wheat. JA was shown to reduce the promoting effect of blue light on seed dormancy in a nitrate-dependent way [96]. Additionally, after ripening promotes expression of JA biosynthesis genes: *TaAllene oxide synthase* (*TaAOS*), *Ta3-ketoacyl coenzyme A* (*TaKAT3*) and *TaLipoxygenase 5* (*TaLOX5*) in wheat grains. However, the level of JA decreases during imbibition (**Table 1**) [72]. The cold-induced release of seed is associated with the increase in JA endogenous content. Cold stratification process promotes the expression of *TaAOS* and *TaAllene oxide cyclase* (*TaAOC*). Furthermore, JA positively regulates *TaNCED1* and *TaNCED2* activity and thus enables seed germination through ABA biosynthesis repression in wheat (**Table 1**) [96, 97].

5.3. Brassinosteroids promote seed germination via repression of ABA signaling

Brassinosteroids (BR) act opposite to ABA signaling in the regulation of seed dormancy and germination. In Arabidopsis, the crucial regulator of seed dormancy, *MFT*, is under BR regulation in seeds. Therefore, MFT acts as a mediator of ABA and BR pathways in seeds [98]. Brassinosteroid insensitive 2 (BIN2) is a GSK3-like kinase playing a negative role in BR signaling, and furthermore, it ensures the communication with ABA signaling. BIN2 interacts with ABI5 and phosphorylates it, which in turn promotes ABI5 activity during seed germination [99]. *TaBIN2* activity is downregulated in the after-ripened wheat seeds (**Table 1**) [100]. Expression analysis also showed the induction of genes encoding the positive components of BR pathway: *TaDE-etiolated 2* (*TaDET2*), *TaDWARF 4* (*TaDWF4*), and *TaBR signaling kinase 2* (*TaBSK2*) in wheat after-ripened grains. *TaDET2* and *TaDWF4* encode crucial enzymes for BR biosynthesis, whereas *TaBSK2* promotes BR signaling (**Table 1**) [100].

5.4. Ethylene represses ABA accumulation and promotes seed dormancy release

Ethylene (ET) is positively related to seed dormancy release and germination promotion. In Arabidopsis, the expression of ET biosynthesis gene, *1-aminocyclopropane-1-carboxylic acid oxidase* (*ACO*), is associated with imbibition; however, cold stratification reduces its expression (**Table 1**) [101]. Ethylene receptors, ethylene triple response 1 (ETR1) and ethylene insensitive 2 (EIN2) play a role in seed dormancy regulation. *etr1* and *ein2* mutants show the increased level of r seed dormancy associated with the increased level of seed ABA content [102, 103]. The higher expression of *NCED3* and lower activation of *CYP707A2* were observed in *ein2* and *etr1* mutants, respectively, compared to the wild type. It suggests a negative role of ethylene in the modulation of ABA pathway in seeds (**Table 1**) [104]. In wheat, after-ripened grains express *TaACO* at a higher level than in dormant grains. Thus, the increased ET content in seeds is associated with dormancy loss also in wheat [100]. The role of ethylene in seed dormancy regulation includes regulation at epigenetic level. *SIN3-like 1* (SNL1) and SNL2 reduce acetylation level of histone 3 lysine 9/18 and histone 3 lysine 14. The double mutant *snl1 snl2* shows reduced seed dormancy together with the increased expression of ethylene biosynthesis genes (*ACO1, ACO4*) and ABA catabolism genes (*CYP707A1, CYP707A2*). Therefore, SNL1 and SNL2 promote seed dormancy through simultaneous modulation of ethylene and ABA content in seeds [105].

6. Conclusions

Proper regulation of seed dormancy is crucial for appropriate timing of germination. Many environmental factors, including light and temperature, exert action on switch from dormancy to germination stage. Their action is mediated by phytohormones: ABA and GA. ABA is a master player for the entrance to and the establishment of seed dormancy. Many ABA-related genes are necessary for the quiescent stage of seeds. Contrary to ABA, GA-mediated pathway promotes germination under favorable conditions. Similar mechanism of seed dormancy regulation exists in monocot plants. The seed response is dependent on the ABA and GA balance.

The ABA-GA crosstalk ensures the precise seed response according to developmental stage, environmental factors, and seasons. Many components of the ABA and GA pathway, for example ABI3, ABI4, ABI5, RGL2, MFT, and DOG1, are responsible for the proper regulation of seed dormancy. Additionally, auxin, jasmonic acid, brassinosteroids, and ethylene modulate the ABA pathway in seeds. Furthermore, epigenetic control of dormancy-related components also occurs. Therefore, seed dormancy regulation appears to be a very elaborate process. In monocot plants, a part of the seed dormancy regulatory mechanism acts in a different manner. Action of MFT and JA pathway seems to be reverse in comparison to dicot plants. A better understanding of precise phytohormonal regulation of seed response of cereals can help in obtaining new varieties with the appropriate seed dormancy level.

Author details

Anna Skubacz and Agata Daszkowska-Golec*

*Address all correspondence to: agata.daszkowska@us.edu.pl

Department of Genetics, Faculty of Biology and Environmental Protection, University of Silesia in Katowice, Katowice, Poland

References

[1] Finch-Savage WE, Leubner-Metzger G. Seed dormancy and the control of germination. New Phytologist. 2006;**171**:501-523. DOI: 10.1111/j.1469-8137.2006.01787.x

[2] Finkelstein R, Reeves W, Ariizumi T, Steber C. Molecular aspects of seed dormancy. Plant Biology. 2008;**59**:387. DOI: 10.1146/annurev.arplant.59.032607.092740

[3] Graeber K, Nakabayashi K, Miatton E, Leubner-Metzger G, Soppe WJJ. Molecular mechanisms of seed dormancy. Plant, Cell & Environment. 2012;**35**:1769-1786. DOI: 10.1111/j.1365-3040.2012.02542.x

[4] Shu K, Liu XD, Xie Q, He ZH. Two faces of one seed: Hormonal regulation of dormancy and germination. Molecular Plant. 2016;**9**:34-45. DOI: 10.1016/j.molp.2015.08.010

[5] Holdsworth MJ, Bentsink L, Soppe WJJ. Molecular networks regulating Arabidopsis seed maturation, after-ripening, dormancy and germination. New Phytologist. 2008;**179**:33-54. DOI: 10.1111/j.1469-8137.2008.02437.x

[6] Gao F, Ayele BT. Functional genomics of seed dormancy in wheat: Advances and prospects. Frontiers in Plant Science. 2014;**5**:458. DOI: 10.3389/fpls.2014.00458

[7] Rodríguez MV, Barrero JM, Corbineau F, Gubler F, Benech-Arnold RL. Dormancy in cereals (not too much, not so little): About the mechanisms behind this trait. Seed Science Research. 2015;**25**:99-119. DOI: 10.1017/S0960258515000021

[8] Liu Y, Shi L, Ye N, Liu R, Jia W, Zhang J. Nitric oxide-induced rapid decrease of abscisic acid concentration is required in breaking seed dormancy in Arabidopsis. New Phytologist. 2009;**183**:1030-1042. DOI: 10.1111/j.1469-8137.2009.02899.x

[9] Matakiadis T, Alboresi A, Jikumaru Y, Tatematsu K, Pichon O, Renou JP, Yuji Kamiya Y, Nambara E, Truong HN. The Arabidopsis abscisic acid catabolic gene CYP707A2 plays a key role in nitrate control of seed dormancy. Plant Physiology. 2009;**149**:949-960. DOI: 10.1104/pp.108.126938

[10] Footitt S, Douterelo-Soler I, Clay H, Finch-Savage WE. Dormancy cycling in Arabidopsis seeds is controlled by seasonally distinct hormone-signaling pathways. Proceedings of the National Academy of Sciences. 2011;**108**:20236-20241. DOI: 10.1073/pnas.1116325108

[11] Shu K, Meng YJ, Shuai HW, Liu WG, Du JB, Liu J, Yang WY. Dormancy and germination: How does the crop seed decide? Plant Biology. 2015;**17**:1104-1112. DOI: 10.1111/plb.12356

[12] Kucera B, Cohn MA, Leubner-Metzger G. Plant hormone interactions during seed dormancy release and germination. Seed Science Research. 2005;**15**:281-307. DOI: 10.1079/SSR2005218

[13] Nambara E, Marion-Poll A. ABA action and interactions in seeds. Trends in Plant Science. 2003;**8**:213-217. DOI: 10.1016/S1360-1385(03)00060-8

[14] Kanno Y, Jikumaru Y, Hanada A, Nambara E, Abrams SR, Kamiya Y, Seo M. Comprehensive hormone profiling in developing Arabidopsis seeds: Examination of the site of ABA biosynthesis, ABA transport and hormone interactions. Plant and Cell Physiology. 2010;**51**:1988-2001. DOI: 10.1093/pcp/pcq158

[15] Mehrotra R, Bhalothia P, Bansal P, Basantani MK, Bharti V, Mehrotra S. Abscisic acid and abiotic stress tolerance—different tiers of regulation. Journal of Plant Physiology. 2014;**171**:486-496. DOI: 10.1016/j.jplph.2013.12.007

[16] Sah SK, Reddy KR, Li J. Abscisic acid and abiotic stress tolerance in crop plants. Frontiers in Plant Science. 2016;**7**:571. DOI: 10.3389/fpls.2016.00571

[17] Lefebvre V, North H, Frey A, Sotta B, Seo M, Okamoto M, Nambara E, Marion-Poll A. Functional analysis of Arabidopsis NCED6 and NCED9 genes indicates that ABA synthesized in the endosperm is involved in the induction of seed dormancy. The Plant Journal. 2006;**45**:309-319. DOI: 10.1111/j.1365-313X.2005.02622.x

[18] Martínez-Andújar C, Ordiz MI, Huang Z, Nonogaki M, Beachy RN, Nonogaki H. Induction of 9-cis-epoxycarotenoid dioxygenase in Arabidopsis thaliana seeds enhances seed dormancy. Proceedings of the National Academy of Sciences. 2011;**108**:17225-17229. DOI: 10.1073/pnas.1112151108

[19] Frey A, Effroy D, Lefebvre V, Seo M, Perreau F, Berger A, Sechet J, To A, North HM, Marion-Poll A. Epoxycarotenoid cleavage by NCED5 fine-tunes ABA accumulation and affects seed dormancy and drought tolerance with other NCED family members. The Plant Journal. 2012;**70**:501-512. DOI: 10.1111/j.1365-313X.2011.04887.x

[20] González-Guzmán M, Abia D, Salinas J, Serrano R, Rodríguez PL. Two new alleles of the abscisic aldehyde oxidase 3 gene reveal its role in abscisic acid biosynthesis in seeds. Plant Physiology. 2004;**135**:325-333. DOI: 10.1104/pp.103.036590

[21] Kushiro T, Okamoto M, Nakabayashi K, Yamagishi K, Kitamura S, Asami T, Nobuhiro Hirai N, Koshiba T, Kamiya Y, Nambara E. The Arabidopsis cytochrome P450 CYP707A encodes ABA 8′-hydroxylases: Key enzymes in ABA catabolism. The EMBO Journal. 2004;**23**:1647-1656. DOI: 10.1038/sj.emboj.7600121

[22] Okamoto M, Kuwahara A, Seo M, Kushiro T, Asami T, Hirai N, Kamiya Y, Koshiba T, Nambara E. CYP707A1 and CYP707A2, which encode abscisic acid 8′-hydroxylases, are indispensable for proper control of seed dormancy and germination in Arabidopsis. Plant Physiology. 2006;**141**:97-107. DOI: 10.1104/pp.106.079475

[23] Matilla AJ, Carrillo-Barral N, del Carmen Rodríguez-Gacio M. An update on the role of NCED and CYP707A ABA metabolism genes in seed dormancy induction and the response to after-ripening and nitrate. Journal of Plant Growth Regulation. 2015;**34**:274-293. DOI: 10.1007/s00344-014-9464-7

[24] Liu Y, Fang J, Xu F, Chu J, Yan C, Schläppi MR, Wang Y, Chu C. Expression patterns of ABA and GA metabolism genes and hormone levels during rice seed development and imbibition: A comparison of dormant and non-dormant rice cultivars. Journal of Genetics and Genomics. 2014;**41**:327-338. DOI: 10.1016/j.jgg.2014.04.004

[25] Chono M, Honda I, Shinoda S, Kushiro T, Kamiya Y, Nambara E, Kawakami N, Kaneko S, Watanabe Y. Field studies on the regulation of abscisic acid content and germinability during grain development of barley: Molecular and chemical analysis of pre-harvest sprouting. Journal of Experimental Botany. 2006;**57**:2421-2434. DOI: 10.1093/jxb/erj215

[26] Leymarie J, Robayo-Romero ME, Gendreau E, Benech-Arnold RL, Corbineau F. Involvement of ABA in induction of secondary dormancy in barley (*Hordeum vulgare* L.) seeds. Plant and Cell Physiology. 2008;**49**:1830-1838. DOI: 10.1093/pcp/pcn164

[27] Sreenivasulu N, Radchuk V, Alawady A, Borisjuk L, Weier D, Staroske N, Fuchs J, Miersch O, Strickert M, Usadel B, Wobus U, Grimm B, Weber H, Weschke W. De-regulation of abscisic acid contents causes abnormal endosperm development in the barley mutant *seg8*. The Plant Journal. 2010;**64**:589-603. DOI: 10.1111/j.1365-313X.2010.04350.x

[28] Mendiondo GM, Leymarie J, Farrant JM, Corbineau F, Benech-Arnold RL. Differential expression of abscisic acid metabolism and signalling genes induced by seed-covering structures or hypoxia in barley (*Hordeum vulgare* L.) grains. Seed Science Research. 2010;**20**: 69-77. DOI: 10.1017/S0960258509990262

[29] Hoang HH, Bailly C, Corbineau F, Leymarie J. Induction of secondary dormancy by hypoxia in barley grains and its hormonal regulation. Journal of Experimental Botany. 2013a;**64**:2017-2025. DOI: 10.1093/jxb/ert062

[30] Millar AA, Jacobsen JV, Ross JJ, Helliwell CA, Poole AT, Scofield G, Reid J, Gubler F. Seed dormancy and ABA metabolism in Arabidopsis and barley: The role of ABA 8′-hydroxylase. The Plant Journal. 2006;**45**:942-954. DOI: 10.1111/j.1365-313X.2006. 02659.x

[31] Gubler F, Hughes T, Waterhouse P, Jacobsen J. Regulation of dormancy in barley by blue light and after-ripening: Effects on abscisic acid and gibberellin metabolism. Plant Physiology. 2008;**147**:886-896. DOI: 10.1104/pp.107.115469

[32] Barrero JM, Talbot MJ, White RG, Jacobsen JV, Gubler F. Anatomical and transcriptomic studies of the coleorhiza reveal the importance of this tissue in regulating dormancy in barley. Plant Physiology. 2009;**150**:1006-1021. DOI: 10.1104/pp.109.137901

[33] Barrero JM, Jacobsen JV, Talbot MJ, White RG, Swain SM, Garvin DF, Gubler F. Grain dormancy and light quality effects on germination in the model grass *Brachypodium distachyon*. New Phytologist. 2012;**193**:376-386. DOI: 10.1111/j.1469-8137.2011.03938.x

[34] Nakashima K, Yamaguchi-Shinozaki K. ABA signaling in stress-response and seed development. Plant Cell Reports. 2013;**32**:959-970. DOI: 10.1007/s00299-013-1418-1

[35] Yoshida T, Mogami J, Yamaguchi-Shinozaki K. ABA-dependent and ABA-independent signaling in response to osmotic stress in plants. Current Opinion in Plant Biology. 2014;**21**:133-139. DOI: 10.1016/j.pbi.2014.07.009

[36] Daszkowska-Golec A. The role of abscisic acid in drought stress: How ABA helps plants to cope with drought stress. Hossain MA, Wani SH, Bhattachajee S, Burritt DJ, Tran LSP, editors. In: Drought Stress Tolerance in Plants. Vol. 2. Cham: Springer International Publishing; 2016. pp. 123-151. DOI: 10.1007/978-3-319-32423-4_5

[37] Umezawa T, Sugiyama N, Mizoguchi M, Hayashi S, Myouga F, Yamaguchi-Shinozaki K, Ishihama Y, Hirayama T, Shinozaki K. Type 2C protein phosphatases directly regulate abscisic acid-activated protein kinases in Arabidopsis. Proceedings of the National Academy of Sciences. 2009;**106**:17588-17593. DOI: 10.1073/pnas.0907095106

[38] Karssen CM, Hilhorst HWM, Koornneef M. The benefit of biosynthesis and response mutants to the study of the role of abscisic acid in plants. In: Pharis RP, Rood SB, editors. Plant Growth Substances 1988. Berlin, Heidelberg: Springer; 1990. pp. 23-31. DOI: 10.1007/978-3-642-74545-4_3

[39] Kim W, Lee Y, Park J, Lee N, Choi G. HONSU, a protein phosphatase 2C, regulates seed dormancy by inhibiting ABA signaling in Arabidopsis. Plant and Cell Physiology. 2013;**54**:555-572. DOI: 10.1093/pcp/pct017

[40] Lee KP, Piskurewicz U, Turečková V, Strnad M, Lopez-Molina L. A seed coat bedding assay shows that RGL2-dependent release of abscisic acid by the endosperm controls embryo growth in Arabidopsis dormant seeds. Proceedings of the National Academy of Sciences. 2010;**107**:19108-19113. DOI: 10.1073/pnas.1012896107

[41] Shu K, Zhang H, Wang S, Chen M, Wu Y, Tang S, Chunyan Liu C, Feng Y, Cao X, Xie Q. ABI4 regulates primary seed dormancy by regulating the biogenesis of abscisic acid and gibberellins in Arabidopsis. PLoS Genetics. 2013;**9**:e1003577. DOI: 10.1371/journal.pgen.1003577

[42] Skubacz A, Daszkowska-Golec A, Szarejko I. The role and regulation of ABI5 (ABA-insensitive 5) in plant development, abiotic stress responses and phytohormone cross-talk. Frontiers in Plant Science. 2016;**7**:1884. DOI: 10.3389/fpls.2016.01884

[43] Parcy F, Valon C, Raynal M, Gaubier-Comella P, Delseny M, Giraudat J. Regulation of gene expression programs during Arabidopsis seed development: Roles of the ABI3 locus and of endogenous abscisic acid. The Plant Cell. 1994;**6**:1567-1582. DOI: 10.1105/tpc.6.11.1567

[44] Parcy F, Valon C, Kohara A, Miséra S, Giraudat J. The ABSCISIC ACID-INSENSITIVE3, FUSCA3, and LEAFY COTYLEDON1 loci act in concert to control multiple aspects of Arabidopsis seed development. The Plant Cell. 1997;**9**:1265-1277. DOI: 10.1105/tpc.9.8.1265

[45] Nambara E, Naito S, McCourt P. A mutant of Arabidopsis which is defective in seed development and storage protein accumulation is a new *abi3* allele. The Plant Journal. 1992;**2**:435-441. DOI: 10.1111/j.1365-313X.1992.00435.x

[46] Ding ZJ, Yan JY, Li GX, Wu ZC, Zhang SQ, Zheng SJ. WRKY41 controls Arabidopsis seed dormancy via direct regulation of ABI3 transcript levels not downstream of ABA. The Plant Journal. 2014;**79**:810-823. DOI: 10.1111/tpj.12597

[47] Wind JJ, Peviani A, Snel B, Hanson J, Smeekens SC. ABI4: Versatile activator and repressor. Trends in Plant Science. 2013;**18**:125-132. DOI: 10.1016/j.tplants.2012.10.004

[48] Lee K, Seo PJ. Coordination of seed dormancy and germination processes by MYB96. Plant Signaling & Behavior. 2015;**10**:e1056423. DOI: 10.1080/15592324.2015.1056423

[49] Yano R, Kanno Y, Jikumaru Y, Nakabayashi K, Kamiya Y, Nambara E. CHOTTO1, a putative double APETALA2 repeat transcription factor, is involved in abscisic acid-mediated repression of gibberellin biosynthesis during seed germination in Arabidopsis. Plant Physiology. 2009;**151**:641-654. DOI: 10.1104/pp.109.142018

[50] Yamagishi K, Tatematsu K, Yano R, Preston J, Kitamura S, Takahashi H, McCourt P, Kamiya Y, Nambara E. CHOTTO1, a double AP2 domain protein of Arabidopsis thaliana, regulates germination and seedling growth under excess supply of glucose and nitrate. Plant and Cell Physiology. 2009;**50**:330-340. DOI: 10.1093/pcp/pcn201

[51] Finkelstein RR. Mutations at two new Arabidopsis ABA response loci are similar to the *abi3* mutations. The Plant Journal. 1994;**5**:765-771. DOI: 10.1046/j.1365-313X.1994.5060765.x

[52] Holman TJ, Jones PD, Russell L, Medhurst A, Tomás SÚ, Talloji P, Marquez J, Schmuths H, Tung S, Taylor I, Footitt S, Bachmair A, Theodoulou FL, Holdsworth MJ. The N-end rule pathway promotes seed germination and establishment through removal of ABA sensitivity in Arabidopsis. Proceedings of the National Academy of Sciences. 2009;**106**:4549-4554. DOI: 10.1073/pnas.0810280106

[53] Cantoro R, Crocco CD, Benech-Arnold RL, Rodríguez MV. In vitro binding of *Sorghum bicolor* transcription factors ABI4 and ABI5 to a conserved region of a *GA 2-OXIDASE* promoter: Possible role of this interaction in the expression of seed dormancy. Journal of Experimental Botany. 2013;**64**:5721-5735. DOI: 10.1093/jxb/ert347

[54] Dekkers BJ, He H, Hanson J, Willems LA, Jamar DC, Cueff G, Rajjou L, Hilhorst HWM, Bentsink L. The Arabidopsis DELAY OF GERMINATION 1 gene affects ABSCISIC ACID INSENSITIVE 5 (ABI5) expression and genetically interacts with ABI3 during Arabidopsis seed development. The Plant Journal. 2016;**85**:451-465. DOI: 10.1111/tpj.13118

[55] Robichaud C, Sussex IM. The response of viviparous-1 and wild type embryos of *Zea mays* to culture in the presence of abscisic acid. Journal of Plant Physiology. 1986;**126**:235-242. DOI: 10.1016/S0176-1617(86)80025-6

[56] Huang T, Qu B, Li HP, Zuo DY, Zhao ZX, Liao YC. A maize viviparous 1 gene increases seed dormancy and preharvest sprouting tolerance in transgenic wheat. Journal of Cereal Science. 2012;**55**:166-173. DOI: 10.1016/j.jcs.2011.11.003

[57] Fan J, Niu X, Wang Y, Ren G, Zhuo T, Yang Y, Lu B, Liu Y. Short, direct repeats (SDRs)-mediated post-transcriptional processing of a transcription factor gene *OsVP1* in rice (*Oryza sativa*). Journal of Experimental Botany. 2007;**58**:3811-3817. DOI: 10.1093/jxb/erm231

[58] Sugimoto K, Takeuchi Y, Ebana K, Miyao A, Hirochika H, Hara N, Ishiyama K, Kobayashi M, Ban Y, Hattori T, Yano M. Molecular cloning of *Sdr4*, a regulator involved in seed dormancy and domestication of rice. Proceedings of the National Academy of Sciences. 2010;**107**:5792-5797. DOI: 10.1073/pnas.0911965107

[59] Rodríguez MV, Mendiondo GM, Maskin L, Gudesblat GE, Iusem ND, Benech-Arnold RL. Expression of ABA signalling genes and ABI5 protein levels in imbibed *Sorghum bicolor* caryopses with contrasting dormancy and at different developmental stages. Annals of Botany. 2009;**104**:975-985. DOI: 10.1093/aob/mcp184

[60] Seo M, Hanada A, Kuwahara A, Endo A, Okamoto M, Yamauchi Y, North H, Marion-Poll A, Sun T, Koshiba T, Kamiya Y, Yamaguchi S, Nambara E. Regulation of hormone metabolism in Arabidopsis seeds: Phytochrome regulation of abscisic acid metabolism and abscisic acid regulation of gibberellin metabolism. The Plant Journal. 2006;**48**:354-366. DOI: 10.1111/j.1365-313X.2006.02881.x

[61] Hoang HH, Sechet J, Bailly C, Leymarie J, Corbineau F. Inhibition of germination of dormant barley (*Hordeum vulgare* L.) grains by blue light as related to oxygen and hormonal regulation. Plant, Cell & Environment. 2014;**37**:1393-1403. DOI: 10.1111/pce.12239

[62] Barrero JM, Downie AB, Xu Q, Gubler F. A role for barley CRYPTOCHROME1 in light regulation of grain dormancy and germination. The Plant Cell. 2014;**26**:1094-1104. DOI: 10.1105/tpc.113.121830

[63] Hofmann N. Cryptochromes and seed dormancy: The molecular mechanism of blue light inhibition of grain germination. The Plant Cell. 2014;**26**:846. DOI: 10.1105/tpc.114.124727

[64] Toh S, Imamura A, Watanabe A, Nakabayashi K, Okamoto M, Jikumaru Y, Hanada A, Aso Y, Ishiyama K, Tamura N, Iuchi S, Kobayashi M, Yamaguchi S, Kamiya Y, Nambara E, Kawakami N. High temperature-induced abscisic acid biosynthesis and its role in the inhibition of gibberellin action in Arabidopsis seeds. Plant Physiology. 2008;**146**:1368-1385. DOI: 10.1104/pp.107.113738

[65] Gibbs DJ, Bacardit J, Bachmair A, Holdsworth MJ. The eukaryotic N-end rule pathway: Conserved mechanisms and diverse functions. Trends in Cell Biology. 2014;**24**:603-611. DOI: 10.1016/j.tcb.2014.05.001

[66] Zheng J, Chen F, Wang Z, Cao H, Li X, Deng X, Soppe WJJ, Li Y, Liu Y. A novel role for histone methyltransferase KYP/SUVH4 in the control of Arabidopsis primary seed dormancy. New Phytologist. 2012;**193**:605-616. DOI: 10.1111/j.1469-8137.2011.03969.x

[67] Zhao M, Yang S, Liu X, Wu K. Arabidopsis histone demethylases LDL1 and LDL2 control primary seed dormancy by regulating DELAY OF GERMINATION 1 and ABA signaling-related genes. Frontiers in Plant Science. 2014;**6**:159. DOI: 10.3389/fpls.2015.00159

[68] Ogawa M, Hanada A, Yamauchi Y, Kuwahara A, Kamiya Y, Yamaguchi S. Gibberellin biosynthesis and response during Arabidopsis seed germination. The Plant Cell. 2003;**15**:1591-1604. DOI: 10.1105/tpc.011650

[69] Yamaguchi S, Kamiya Y, Sun TP. Distinct cell-specific expression patterns of early and late gibberellin biosynthetic genes during Arabidopsis seed germination. The Plant Journal. 2001;**28**:443-453. DOI: 10.1046/j.1365-313X.2001.01168.x

[70] Debeaujon I, Koornneef M. Gibberellin requirement for Arabidopsis seed germination is determined both by testa characteristics and embryonic abscisic acid. Plant Physiology. 2000;**122**:415-424. DOI: 10.1104/pp.122.2.415

[71] Yamauchi Y, Ogawa M, Kuwahara A, Hanada A, Kamiya Y, Yamaguchi S. Activation of gibberellin biosynthesis and response pathways by low temperature during imbibition of *Arabidopsis thaliana* seeds. The Plant Cell. 2004;**16**:367-378. DOI: 10.1105/tpc.018143

[72] Liu A, Gao F, Kanno Y, Jordan MC, Kamiya Y, Seo M, Ayele B. Regulation of wheat seed dormancy by after-ripening is mediated by specific transcriptional switches that induce changes in seed hormone metabolism and signaling. PLoS One. 2013;**8**:e56570. DOI: 10.1371/journal.pone.0056570

[73] Penfield S, Josse EM, Kannangara R, Gilday AD, Halliday KJ, Graham IA. Cold and light control seed germination through the bHLH transcription factor SPATULA. Current Biology. 2005;**15**:1998-2006. DOI: 10.1016/j.cub.2005.11.010

[74] Gabriele S, Rizza A, Martone J, Circelli P, Costantino P, Vittorioso P. The Dof protein DAG1 mediates PIL5 activity on seed germination by negatively regulating GA biosynthetic gene *AtGA3ox1*. The Plant Journal. 2010;**61**:312-323. DOI: 10.1111/j.1365-313X.2009.04055.x

[75] Hoang HH, Sotta B, Gendreau E, Bailly C, Leymarie J, Corbineau F. Water content: A key factor of the induction of secondary dormancy in barley grains as related to ABA metabolism. Physiologia Plantarum. 2013b;**148**:284-296. DOI: 10.1111/j.1399-3054.2012.01710.x

[76] Rodríguez MV, Mendiondo GM, Cantoro R, Auge GA, Luna V, Masciarelli O, Benech-Arnold RL. Expression of seed dormancy in grain sorghum lines with contrasting pre-harvest sprouting behavior involves differential regulation of gibberellin metabolism genes. Plant and Cell Physiology. 2012;**53**:64-80. DOI: 10.1093/pcp/pcr154

[77] Hauvermale AL, Tuttle KM, Takebayashi Y, Seo M, Steber CM. Loss of Arabidopsis thaliana seed dormancy is associated with increased accumulation of the GID1 GA hormone receptors. Plant and Cell Physiology. 2015;**56**:1773-1785. DOI: 10.1093/pcp/pcv084

[78] Tyler L, Thomas SG, Hu J, Dill A, Alonso JM, Ecker JR, Sun TP. DELLA proteins and gibberellin-regulated seed germination and floral development in Arabidopsis. Plant Physiology. 2004;**135**:1008-1019. DOI: 10.1104/pp.104.039578

[79] Steber CM, Cooney SE, McCourt P. Isolation of the GA-response mutant sly1 as a suppressor of ABI1-1 in *Arabidopsis thaliana*. Genetics. 1998;**149**:509-521

[80] Footitt S, Slocombe SP, Larner V, Kurup S, Wu Y, Larson T, Graham I, Baker A, Holdsworth M. Control of germination and lipid mobilization by COMATOSE, the Arabidopsis homologue of human ALDP. The EMBO Journal. 2002;**21**:2912-2922. DOI: 10.1093/emboj/cdf300

[81] Cao D, Hussain A, Cheng H, Peng J. Loss of function of four DELLA genes leads to light- and gibberellin-independent seed germination in Arabidopsis. Planta. 2005;**223**:105-113. DOI: 10.1007/s00425-005-0057-3

[82] Richards DE, King KE, Ait-Ali T, Harberd NP. How gibberellin regulates plant growth and development: A molecular genetic analysis of gibberellin signaling. Annual Review of Plant Biology. 2001;**52**:67-88. DOI: 10.1146/annurev.arplant.52.1.67

[83] Bentsink L, Jowett J, Hanhart CJ, Koornneef M. Cloning of *DOG1*, a quantitative trait locus controlling seed dormancy in Arabidopsis. Proceedings of the National Academy of Sciences. 2006;**103**:17042-17047. DOI: 10.1073/pnas.0607877103

[84] Graeber K, Linkies A, Steinbrecher T, Mummenhoff K, Tarkowská D, Turečková V, Ignatz M, Sperber K, Voegele A, de Jong H, Urbanová T, Strnad M, Leubner-Metzger G. DELAY OF GERMINATION 1 mediates a conserved coat-dormancy mechanism for the temperature-and gibberellin-dependent control of seed germination. Proceedings of the National Academy of Sciences. 2014;**111**:E3571-E3580. DOI: 10.1073/pnas.1403851111

[85] Seo M, Kanno Y, Frey A, North HM, Marion-Poll A. Dissection of Arabidopsis NCED9 promoter regulatory regions reveals a role for ABA synthesized in embryos in the regulation of GA-dependent seed germination. Plant Science. 2016;**246**:91-97. DOI: 10.1016/j.plantsci.2016.02.013

[86] Piskurewicz U, Jikumaru Y, Kinoshita N, Nambara E, Kamiya Y, Lopez-Molina L. The gibberellic acid signaling repressor RGL2 inhibits Arabidopsis seed germination by stimulating abscisic acid synthesis and ABI5 activity. The Plant Cell. 2008;**20**:2729-2745. DOI: 10.1105/tpc.108.061515

[87] Yuan K, Rashotte AM, Wysocka-Diller JW. ABA and GA signaling pathways interact and regulate seed germination and seedling development under salt stress. Acta Physiologiae Plantarum. 2011;**33**:261-271. DOI: 10.1007/s11738-010-0542-6

[88] Liu X, Hu P, Huang M, Tang Y, Li Y, Li L, Hou X. The NF-YC-RGL2 module integrates GA and ABA signalling to regulate seed germination in Arabidopsis. Nature Communications. 2016;**7**:12768. DOI: 10.1038/ncomms12768

[89] Vaistij FE, Gan Y, Penfield S, Gilday AD, Dave A, He Z, Josse E, Choi G, Halliday KJ, Graham IA. Differential control of seed primary dormancy in Arabidopsis ecotypes by the transcription factor SPATULA. Proceedings of the National Academy of Sciences. 2013;**110**:10866-10871. DOI: 10.1073/pnas.1301647110

[90] Ibarra SE, Tognacca RS, Dave A, Graham IA, Sánchez RA, Botto JF. Molecular mechanisms underlying the entrance in secondary dormancy of Arabidopsis seeds. Plant, Cell & Environment. 2016;**39**:213-221. DOI: 10.1111/pce.12607

[91] Xi W, Liu C, Hou X, Yu H. MOTHER OF FT AND TFL1 regulates seed germination through a negative feedback loop modulating ABA signaling in Arabidopsis. The Plant Cell. 2010;**22**:1733-1748. DOI: 10.1105/tpc.109.07 3072

[92] Nakamura S, Abe F, Kawahigashi H, Nakazono K, Tagiri A, Matsumoto T, Utsugi S, Taiichi Ogawa T, Handa H, Ishida H, Mori M, Kawaura K, Ogihara Y, Miura H. A wheat homolog of MOTHER OF FT AND TFL1 acts in the regulation of germination. The Plant Cell. 2011;**23**:3215-3229. DOI: 10.1105/tpc.111.088492

[93] Liu X, Zhang H, Zhao Y, Feng Z, Li Q, Yang HQ, Luan S, Li J, He ZH. Auxin controls seed dormancy through stimulation of abscisic acid signaling by inducing ARF-mediated ABI3 activation in Arabidopsis. Proceedings of the National Academy of Sciences. 2013;**110**:15485-15490. DOI: 10.1073/pnas.1304651110

[94] Preston J, Tatematsu K, Kanno Y, Hobo T, Kimura M, Jikumaru Y, Yano R, Kamiya Y, Nambara E. Temporal expression patterns of hormone metabolism genes during imbibition of Arabidopsis thaliana seeds: A comparative study on dormant and non-dormant accessions. Plant and Cell Physiology. 2009;**50**:1786-1800. DOI: 10.1093/pcp/pcp121

[95] Dave A, Vaistij FE, Gilday AD, Penfield SD, Graham IA. Regulation of Arabidopsis thaliana seed dormancy and germination by 12-oxo-phytodienoic acid. Journal of Experimental Botany. 2016;**67**:2277-2284. DOI: 10.1093/jxb/erw028

[96] Jacobsen JV, Barrero JM, Hughes T, Julkowska M, Taylor JM, Xu Q, Gubler F. Roles for blue light, jasmonate and nitric oxide in the regulation of dormancy and germination in wheat grain (*Triticum aestivum* L.). Planta. 2013;**238**:121-138. DOI: 10.1007/s00425-013-1878-0

[97] Xu Q, Truong TT, Barrero JM, Jacobsen JV, Hocart CH, Gubler F. A role for jasmonates in the release of dormancy by cold stratification in wheat. Journal of Experimental Botany. 2016;**67**:3497-3508. DOI: 10.1093/jxb/erw172

[98] Xi W, Yu H. MOTHER OF FT AND TFL1 regulates seed germination and fertility relevant to the brassinosteroid signaling pathway. Plant Signaling & Behavior. 2010;**5**:1315-1317. DOI: 10.4161/psb.5.10.13161

[99] Hu Y, Yu D. BRASSINOSTEROID INSENSITIVE2 interacts with ABSCISIC ACID INSENSITIVE5 to mediate the antagonism of brassinosteroids to abscisic acid during seed germination in Arabidopsis. The Plant Cell. 2014;**26**:4394-4408. DOI: 10.1105/tpc.114.130849

[100] Chitnis VR, Gao F, Yao Z, Jordan MC, Park S, Ayele BT. After-ripening induced transcriptional changes of hormonal genes in wheat seeds: The cases of brassinosteroids, ethylene, cytokinin and salicylic acid. PloS One. 2014;**9**:e87543. DOI: 10.1371/journal.pone.0087543

[101] Narsai R, Law SR, Carrie C, Xu L, Whelan J. In-depth temporal transcriptome profiling reveals a crucial developmental switch with roles for RNA processing and organelle metabolism that are essential for germination in Arabidopsis. Plant Physiology. 2011;**157**:1342-1362. DOI: 10.1104/pp.111.183129

[102] Beaudoin N, Serizet C, Gosti F, Giraudat J. Interactions between abscisic acid and ethylene signaling cascades. The Plant Cell. 2000;**12**:1103-1115. DOI: 10.1105/tpc.12.7.1103

[103] Chiwocha SD, Cutler AJ, Abrams SR, Ambrose SJ, Yang J, Ross AR, Kermode AR. The *etr1-2* mutation in Arabidopsis thaliana affects the abscisic acid, auxin, cytokinin and gibberellin metabolic pathways during maintenance of seed dormancy, moist-chilling and germination. The Plant Journal. 2005;**42**:35-48. DOI: 10.1111/j.1365-313X.2005.02359.x

[104] Cheng WH, Chiang MH, Hwang SG, Lin PC. Antagonism between abscisic acid and ethylene in Arabidopsis acts in parallel with the reciprocal regulation of their metabolism and signaling pathways. Plant Molecular Biology. 2009;**71**:61-80. DOI: 10.1007/s11103-009-9509-7

[105] Wang Z, Cao H, Sun Y, Li X, Chen F, Carles A, Li Y, Ding M, Zhang C, Deng X, Soppe WJ, Yong-Xiu Liu YX. Arabidopsis paired amphipathic helix proteins SNL1 and SNL2 redundantly regulate primary seed dormancy via abscisic acid-ethylene antagonism mediated by histone deacetylation. The Plant Cell. 2013;**25**:149-166. DOI: 10.1105/tpc.112.108191

Chapter 5

Strigolactone Signaling in Plants

Marek Marzec

Additional information is available at the end of the chapter

http://dx.doi.org/10.5772/intechopen.68497

Abstract

Strigolactones (SLs) are a new group of recently described phytohormones. They were found to be involved in the communication between plant roots and symbiotic bacteria or fungi, but also in the interactions between roots of host plants and germinating seeds of parasitic plants. Over the years, however, it has become clear that SLs play a regulatory role in many aspects of plant growth and development. Extensive studies on plant model species *Arabidopsis thaliana* L. and *Oryza sativa* L. have uncovered the molecular mechanisms of SL biosynthesis and signaling. In some aspects, the SL perception and signaling correspond to the already known mechanisms described for other phytohormones, but in other points, they seem to be unique in the plant kingdom. This chapter summarizes the recent discoveries in the signal transduction pathway of SLs and describes the model of SL perception and signaling.

Keywords: strigolactones (SLs), perception, signaling, degradation, SCF complex, receptor, repressor

1. Introduction

Strigolactones (SLs) are a class of carotenoid derivates. They were first discovered in root exudates of cotton and found germinating to be potent stimulants for seed germination of the parasitic plant *Striga lutea* Lour. (witchweed) [1], for review, see [2]. Why plants should produce and secrete a signal molecule that is recognized by its parasites was revealed much later when SLs were identified as signal molecules involved in the establishment and maintenance of interactions with arbuscular mycorrhizal fungi (AMF) [3] and N-fixing bacteria [4]. This also marked the time when SLs were classified as a new class of phytohormones, based on the semi-dwarf and shoot-branched mutants of

Arabidopsis thaliana L. and *Oryza sativa* L. [5, 6]. In the following years, it was discovered that SLs are involved in additional aspects of plant development (**Figure 1**), that is, in the regulation of root system development by promotion of primary roots (PRs) elongation and inhibition of adventitious roots (ARs) formation. The SL effect on lateral roots (LRs) development was found to depend on the availability of nutrients, especially phosphorous (P) and nitrogen (N). Under optimal growth conditions, SLs will inhibit the elongation of LRs, but when plants are exposed to starvation stress SLs induce LR growth, for review see [7]. It thus became clear that SLs play a role in the complex plant response to nutrient stresses. Under both P and N deficiency conditions, the synthesis of SLs is increased and larger amounts of this hormone are secreted into the soil, probably to promote the symbiotic relations with AMF and bacteria. The elevated levels of SLs also influence the plant phenotype by suppressing shoot growth and stimulating root development, thus adapting the plant to starvation conditions, for review see [8, 9]. The contribution of SLs in plant adaptation to the other stresses such as drought is

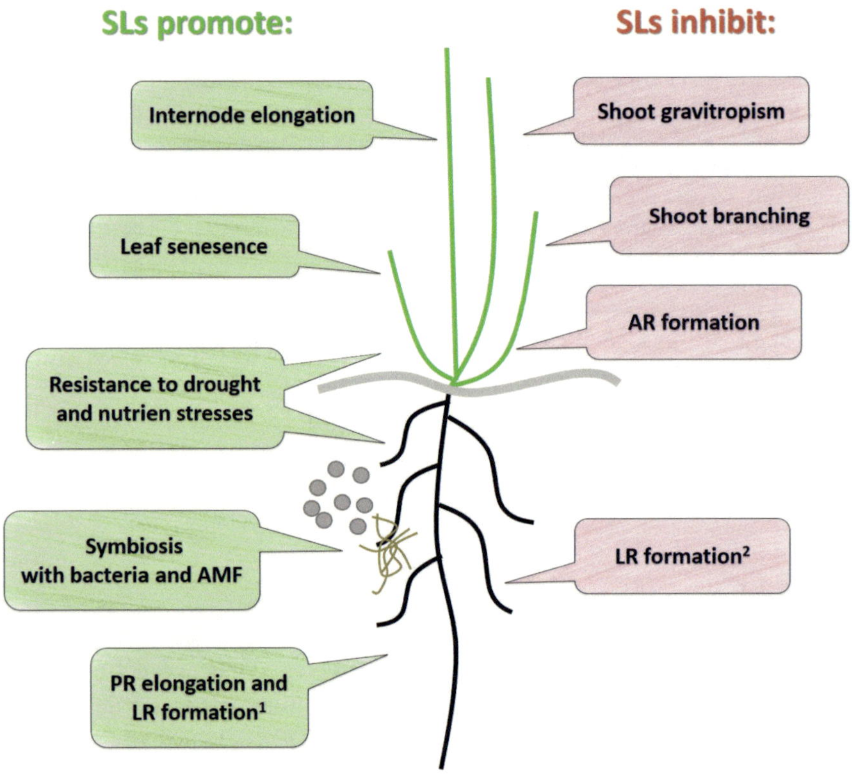

Figure 1. SLs regulate plant development by the promotion of internode elongation, leaf senescence, elongation of primary root (PR) and lateral root (LR)[1] or inhibition of shoot branching, shoot gravitropism, and formation of adventitious root (AR) and LR[2]. Additionally, SLs promote the symbiosis with arbuscular mycorrhizal fungi (AMF) and N-fixing bacteria, and play a role in plant adaptation to drought and nutrient stresses. [1]SLs promote LR elongation under starvation stress and [2]inhibit LR elongation under optimal growth conditions.

unclear. According to some recent studies, SL mutants of *A. thaliana* are more sensitive to drought stress in comparison to the wild-type plants [10, 11]. The results were not conclusive, however. In one study, the SL-signaling mutants, *max2-1* and *max2-2*, were found to be hypersensitive to drought, whereas the SL biosynthesis mutants, *max1*, *max3* and *max4*, were not [10]. In a second study, both groups of mutants, defective in SL-biosynthesis (*max3-11*, *max4-7*) or SL signaling (*max2-3*), were more sensitive to drought [11]. Additional studies on *Lotus japonicas* L. [12] and *Solanum lycopersicum* L. [13, 14] confirmed that SLs together with abscisic acid (ABA) play a role in plant adaptation to the limited water conditions. In response to drought, tomato plants show decreased SL biosynthesis in the roots but increased biosynthesis in the shoots [14]. Results like these feed the impression that SLs may well present a broad-spectrum class of phytohormones (**Figure 1**). Based on an *in silico* analysis of the genes involved in SL biosynthesis in *A. thaliana* and rice, it has been postulated that SLs may also participate in pathogen defense mechanisms and plant responses to wounding, cold stress or flooding [15]. Until now, the experimental evidences have confirmed the role of SLs in plant resistance to bacterial and fungal pathogens, reviewed by Marzec [16].

Up to now, more than 20 naturally occurring SLs, synthetized from the carlactone precursor, have been identified in the plant kingdom [17]. They share a similar structure, composed of a tricyclic lactone (ABC rings) connected to a butenolide group (D ring) by an enol-ether bond (**Figure 2**). SLs are divided into two groups based on the stereochemical differences at the junctions between B and C rings: the orobanchol group with an α-oriented C ring and the strigol group with a β-oriented C ring (**Figure 2**) [18]. SLs are mainly produced in roots and transported to the shoot *via* specific transporters [19]. Alternatively, they might be also produced in the above-ground parts of plants [20]. SLs biosynthesis started in plastids with the conversion of all-*trans*-β-carotene into 9-*cis*-β-carotene by the carotenoid isomerase Dwarf27 (D27), an iron-containing protein [21–23]. The following stages of SLs biosynthesis are conducted by the carotenoid cleavage dioxygenases (CCDs) CCD7 [24, 25] and CCD8 [26, 27] and first involve the transformation of 9-*cis*-β-carotene into 9-*cis*-β-apo-10′-carotenal which is subsequently converted into carlactone [23]. Carlactone, a precursor for all known SLs, has no activity attributed to SLs, until it is converted into carlatonic acid by more axillary growth (MAX1) that belongs to the cytochrome P450 family [23]. The final methylation of carlatonic acid is mediated by an as-yet unknown enzyme [28] (**Figure 3**).

Whereas the *A. thaliana* genome contains only one MAX1 gene, different rice varieties were characterized from two to five MAX1 homologs, which are involved in the synthesis of different SLs [29]. Still the open question remains if *A. thaliana* MAX1 mediates in the production of all SLs or only in the specific ones. The enzyme involved in the subsequent steps of SLs specialization, lateral branching oxidoreductase (LBO), converts methylated carlatonic acid into an as-yet unidentified strigolactone-like compound [30]. Characterization of the product of LBO activity and identification of additional enzymes involved in the last stages of SLs biosynthesis will be essential to compare the production pathways of this hormone in mono- and dicot species.

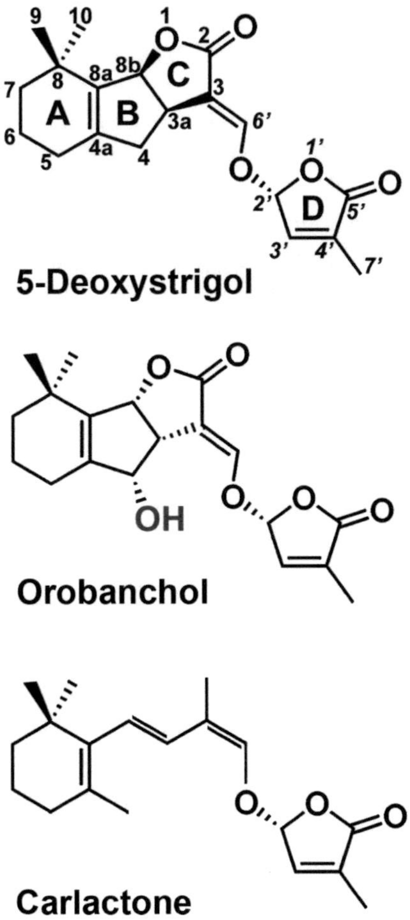

Figure 2. Structures of SL precursor carlactone and SLs represent two main stereochemical groups: strigol-type SL with a β-oriented C ring – 5-deoxystrigol and orobanchol type with an α-oriented C ring – orobanchol. Differences are present at the 8b and 3a positions between B and C rings.

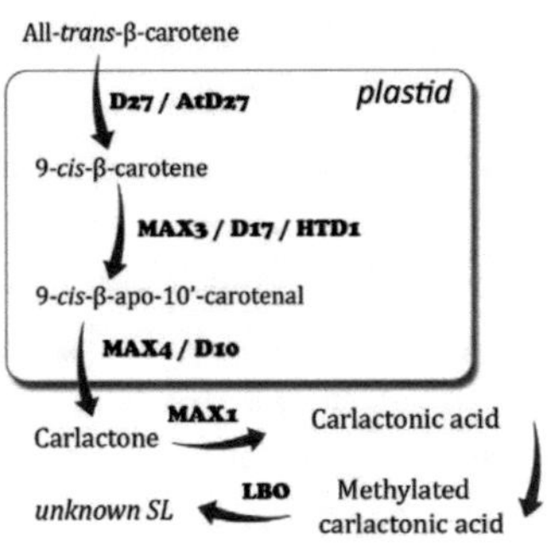

Enzyme	*A. thaliana*	Rice
D27	AtD27 (At1g03055)	D27 (Os11g0587000)
CCD7	MAX3 (At2g42620)	D17/HTD1 (Os04g0550600)
CCD8	MAX4 (At4g32810)	D10 (Os01g0746400)
MAX1	MAX1 (At2g26170)	OsMAX1 (Os1g0700900; Os01g0701500 Os01g0701400; Os02g0221900 Os06g0565100)
LBO	LBO (At3g21420)	

Figure 3. Scheme of SLs biosynthesis and a list of enzymes involved in this process. Descriptions are given in the text.

2. Perception of SLs

The Dwarf14 (D14) protein is the only known receptor of SLs. This protein was originally identified in rice [31] and later on found in other species, such as *A. thaliana* (AtD14) [32], petunia (DAD2) [33], *Hordeum vulgare* L. (HvD14) [34] or *Populus trichocarpa* Torr. & A. Gray (PtD14) [35]. All D14 proteins belong to the α/β-hydrolase family and exhibit enzymatic activity, which is unusual for hormone receptors. D14 proteins can not only bind SL molecule but also hydrolyze the ligand, which is crucial for the next steps of the signaling cascade [36]. The entry to the active site pocket of D14 is surrounded by four helicases, and mutation in which the size of this aperture is reduced causes an insensitivity to SLs (**Figure 4**) [34]. When a ligand is docked to the receptor, a nucleophilic attack separates the ABC part of the SL molecule from the D ring [37]. The hydrolyze activity of D14 depends on the presence of a highly conserved catalytic Ser/His/Asp triad. Replacement of these amino acids results in a loss of D14 activity and sensitivity to SLs [33]. It was also shown

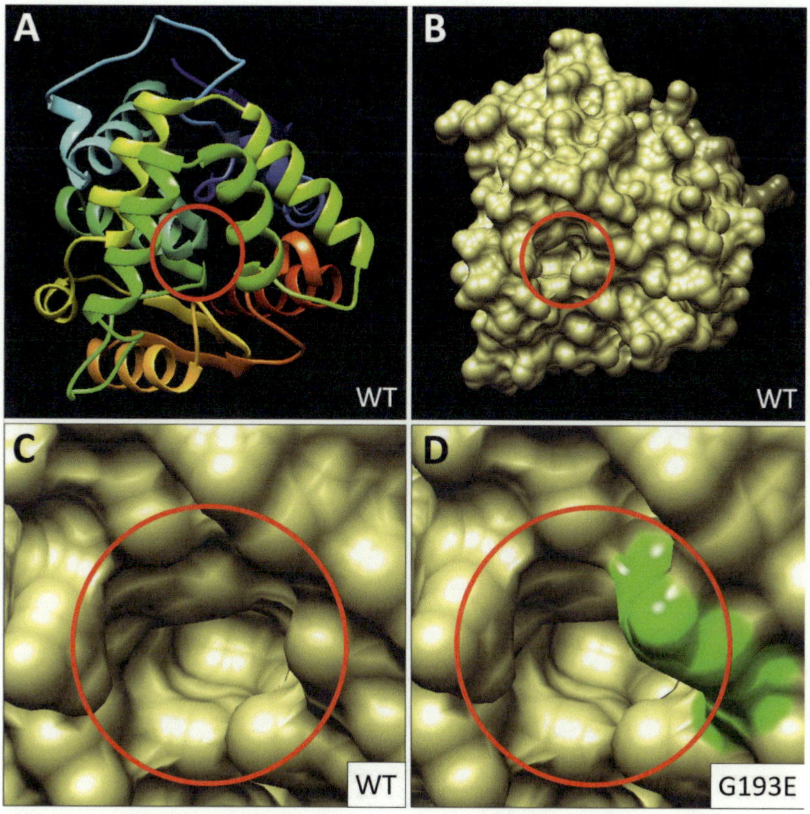

Figure 4. Visualization of SL receptor HvD14. (A) Structure of HvD14. Circle indicates entry to the active site pocket, surrounded by four helices. (B) 3D structure visualization of HvD14. (C) Detailed view of the entry to the active site pocket of the wild-type protein and (D) mutated protein with smaller aperture, resulting in insensitivity to SLs.

that serine from the catalytic triad is involved in docking of SLs into D14 [38]. With an average rate of 0.3 molecule/min, D14-mediated hydrolysis of SLs into non-active derivatives is very slow, indicating that this is not the main function of D14 [33, 36]. Crystallographic analysis indicates that the degradation of SL molecules by D14 brings about a change in receptor conformation, which is necessary for the interaction between D14 and other components from SL-signaling pathway [39]. After nucleophilic attack and release of the ABC part, the D ring remains within the receptor that now assumes a "closed" state unable to bind further molecules, reviewed by Waters [7]. This change in conformation destabilizes the D14 receptor, thus initiating its own degradation [40] (**Figure 5**). This is the first known case where hormone hydrolysis by a receptor causes the degradation receptor as well.

Since SLs are involved in the regulation of the development of different organs, it was expected that D14 will be located in almost all plant tissues. Expression analysis of *Atd14* in *A. thaliana*, however, showed markedly higher levels in vascular tissues of roots and shoots [41]. The discrepancy between expression and distribution pattern of the D14 protein was explained, when the intercellular transport of D14 *via* the phloem was uncovered. This transport is SL-independent and also in the SL biosynthesis mutants it was observed that D14 was delivered into axillary buds [42] leaving the question by what mechanism D14 is transported and how it is been used by plants to regulate the development and adaptation to different stresses.

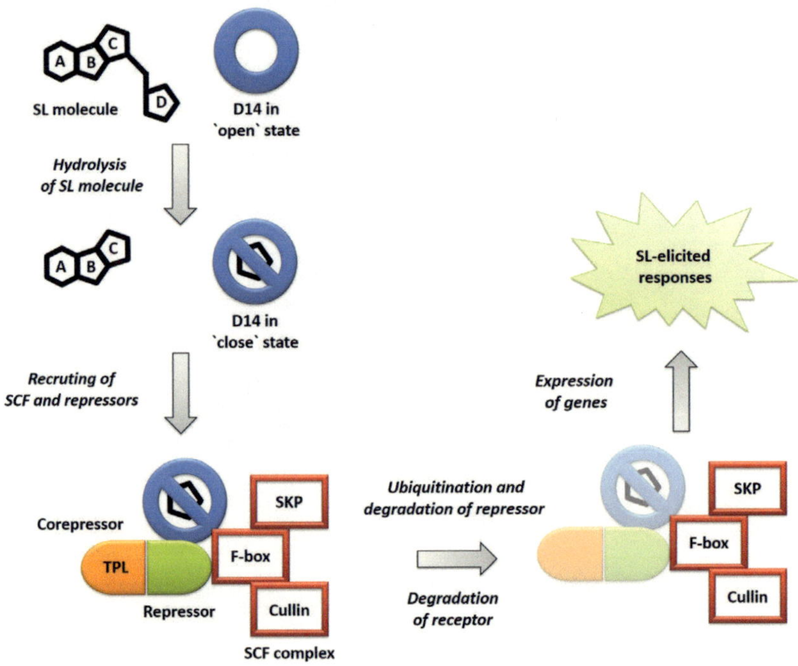

Figure 5. Overview of SL-signaling cascade including hydrolysis of SL molecules by receptor and change of the receptor conformation, which allows the interaction with the SCF complex and repressor. Ubiquitination of the repressor, mediated by the SCF complex, results in the expression of genes from the TCP family.

It has to be highlighted that D14 protein is specific receptor only for SLs. There are close D14 homologs, such as receptor for Karrikins (KARs): Karrikin-Insensitive2 (KAI2) that also belongs to the α/β-hydrolase. However, it was experimentally confirmed that D14 and KAI2 exhibit the different ligand specification [32].

3. Transduction of SL signal

The common mechanism for transducing phytohormone signals is the degradation of proteins called repressors. This degradation is mediated by a SKP1-Cullin-F-box complex (SCF), composed of three proteins that upon binding to repressors mark them for proteasomal degradation *via* ubiquitination. Cullin is a main structural part of the SCF complex controlling the connection of the whole complex to the ligase E3. The SKP1 component is responsible for the binding of the specific F-box protein which designates the protein for degradation [43]. In theory, the F-box protein renders specificity to the whole CSF complex and should be specific for different hormones. In practice, however, evidence shows that other components interacting with the SCF complex can influence its specificity.

An F-box protein which was part of an SCF complex and involved in SL signaling was identified in the *A. thaliana*-mutant *max2* and the rice-mutant *d3* that were insensitive to treatment with SLs [44, 45]. For the long time that mutants were used to describe the role of SLs in the plant growth and development, recently MAX2/D3 has been found to be involved in both SL- and KAR-signaling pathways, reviewed by Waters [7]. For this reason, the phenotype of *max2/d3* mutants cannot be directly linked to the function of SLs. Evidence showing that MAX2-guided degradation of transcription factors that is dependent also on the other phytohormone class of brassinosteroids (BRs) indicates that this F-box protein fulfills a wide range of this F-box protein in hormone signaling [46].

In *A. thaliana,* MAX2 interacts with AtCullin1 and Arabidopsis Serine/Threonine Kinase1 (ASK1) [47], whereas in rice D3 is part of a SCF complex together with OsCullin1 and *O. sativa* SKP1-Like 1/5/20 (OSK 1/5/20) [48]. Although these complexes indicate a conserved mechanism for SL signal transduction in both mono- and dicots, the presence of three different OSKs in rice may suggest that complexes with OSK1/5/20 recognize different substrates and are involved in different SL-dependent processes [48]. MAX2/D3 show nuclear localizations and SL-dependent interactions between F-box protein D3 with SL receptor D14 were reported in rice. Obtained results indicate that this interaction is mediated by the presence of SLs, and it depends on the concentration of SLs and it is also stereoisomer-specific. In a "closed" state, D14 is able to interact with D3 (**Figure 5**), while the version of D14 with mutations in or near the active pocket site shown reduced interactions with D3. This suggests that after SL-mediated changed conformation of D14, D3 can bind D14 close to the active pocket side entry [40].

Based on the similarity of MAX2/D3 protein to other hormone receptors such as jasmonate receptor Coronatine Insensitive1 (COI1) [49] or auxin receptor Transport Inhibitor Response1 (TIR1) [50], it was predicted that MAX2/D3 may also be involved in SL perception. Although there is no evidence that MAX2/D3 can interact with SLs, there are reasons to assume that MAX2/D3 may act as a receptor for other signaling molecules.

For a long time, it was not known which proteins are recognized by the $SCF^{MAX2/D3}$ complex, but recently the SL repressors degraded during SL signal transduction were identified in rice (D53) [51, 52] and in *A. thaliana* (Suppressor of MAX2-Like 6 to 8, SMXL 6 to 8) [53–55]. Gain-of-function mutation in D53 resulted in semi-dwarf plants with increased tillering, a phenotype which is characteristic for SL mutants. Similar effects were observed after overexpression of *D53*, whereas reduced expression of *D53* in a *d53*-mutant background inhibited tiller formation [52]. D53 shows the nuclear localization and it was confirmed that in the presence of SL molecules degradation of D53 occurs. This degradation proceeds through the proteasome-dependent pathway and requires the presence of D14 and D3 proteins [51, 52]. There are evidences that D53 may also interact with D3 in the absence of D14, although this interaction is less efficient. In contrast to rice, where only one SL repressor has been identified, *A. thaliana* contains three proteins—SMXL 6 to 8—that may act redundantly. First reports indicated that only a triple mutant *smxl6/7/8* will result in a phenotype with reduced number of tillers [54]. Later on, it was found that the phenotype characteristic for SL mutants could be produced by the expression of the non-degradable form of SMXL 7 under a native promoter [55]. All three SMXLs interact with D14 and are degraded in an SL-dependent manner in the presence of MAX2 and D14 [53, 54]. It still remains an open question whether SMXLs 6 to 8 do act redundantly or they are involved in different responses to SLs.

The SL repressors of both *A. thaliana* and rice contain the conserved amino acid sequences (F/L-D-L-N-L) which is known as an ethylene-responsive element-binding factor-associated amphiphilic repression (EAR) motif. This motif plays a key role in interactions with transcriptional corepressors from Topless (TPL) and Topless-Related Proteins (TPR) families [52, 53]. TPL and TPR regulate the expression of genes in response to different classes of hormones, such as auxin or jasmonates [56]. The presence of an EAR motif in SL repressors suggests that D53/SMXLs may bind TPL/TPR corepressors. An ensuing degradation of these corepressors may then result in the expression of transcriptional factors, previously suppressed by TPL/TRP. Interactions between SMXL6 to 8 and proteins from the TPR family were already confirmed using a yeast-two hybrid and co-immunoprecipitation assays [54].

4. Transcription response to SL signal

Phytohormones induce a change in gene expression. This response is usually mediated by transcription factors. Until now, only one family of transcription factors has been identified as a downstream component in SL signaling, namely the TEOSINTE BRANCHED1/CYCLOIDEA/PROLIFERATING CELL FACTOR1 family (TCP). In different species, single transcription factors from this family, related to SL signal, were already characterized: Branched1 (BRC1) in *A. thaliana* [57], Fine Culm1/Teosinte Branched1 (FC1/OsTB1) in rice [58], TB1 in *Zea mays* L. [59] and PsBRC1 in *Pisum sativum* L. [60]. Expression of these genes is particularly strong in axillary buds, and mutations in these genes lead to an increased branched phenotype, which cannot be reverse by treatment with SLs. Elevated levels of *AtBRC1* expression were found after SL treatment and in the triple *smxl6/7/8* mutant, whereas decreased levels were found in SL biosynthesis mutants [53, 54]. Similar results were obtained for the *AtBRC1* target gene *Homeobox-Leucine Zipper Protein 53* (*HB53*) [54], confirming that the transcription factors from TCP family act as elicitors in the SL-signaling cascade. There remain, however, some differences between mono- and dicot species. In *A. thaliana* and pea, the expression of *AtBRC1/PsBRC1* is upregulated by SL treatment [60], whereas in rice and maize, the expression of *FC1/OsTB1* and *TB1* is not elevated by SL treatment [61, 62]. This difference is sometimes explained by assuming that monocots contain additional, still unknown, transcription factors that may act as *AtBRC1/PsBRC1* in dicots.

5. SL versus KAR signaling

The unique features of SLs signaling have been discussed elsewhere [7, 63]. Here, the similarities and differences between SLs and KARs will be summarized. Though SLs and KARs play different roles in plant development [64], there are some striking similarities in the signal transduction mechanisms of these two classes of plant growth regulators [65] what might be crucial in understanding the mechanisms of their actions in plants. As already mentioned, the signal transduction of SL and KAR is mediated by the same F-box protein MAX2. However, since the signals generated by SLs and KARs are not interchangeable these phytohormones must be recognized by different receptors. Indeed, the D14 receptor has found to be specific for SLs, whereas the KAI2 receptor is specific for KARs and based on the differences in the size of active pocket site they cannot recognize the signal from second group of plant growth regulators [66, 67]. Both D14 and KA12 display a conserved catalytic triad, but only in case of D14 its catalytic activity was confirmed [33, 37]. Not only has a catalytic function of KAI2 never been proven, modeling studies of the KAR-KAI2 complex indicate that the distance between the KAR molecule and the catalytic Ser from KAI2 prohibits nucleophilic attack [67, 68]. Nevertheless, since mutation in the catalytic triad of KA12 can abolish the function of this receptor [69], the catalytic triad of the KA12 receptor may be essential for ligand binding. Similar observations have been made for D14 [38]. The second similarity between both receptors is their degradation during perception, though in the case of KAI2 the presence of MAX2 is not required for its degradation [70].

MAX2 is a component of SCF complexes which are involved in the conversion of SL and KAR signals. Therefore, the phenotypic effects caused by a mutation of MAX2 are due to an insensitivity to both plant growth regulators.

Due to the presence of different receptors, the respective SCF complexes guide the degradation of different suppressors: SMXL6 to 8 in the case of D14-SL-Max2 and Supressor of Max2 1 (SMAX1) in the case of KAI2-KAR-MAX2 [71]. *SMAX1* and *SMXL6/7/8* show similar patterns of expression in *A. thaliana* seedlings and all four proteins may interact with TPL corepressors *via the* EAR motif [53] which indicates that ultimately SLs and KARs may regulate gene expression by a similar mechanism.

6. Conclusions

Since their classification as phytohormones, great progress has been made uncovering the mechanisms of SL signaling, and identifying the main components of the SL signal transduction pathway in both mono- and dicots. Certain aspects of SL perception have been found to be unique among plant hormones, requiring additional research to understand these phenomena in more detail. SLs share a number of the signaling components with the KARs group of plant growth regulators. Attention should also be paid to the respective receptor molecules since they represent the crucial element separating both signal cascades. Presently, our knowledge about the transcriptional responses to treatment with SLs and KARs is limited and information on the targets of SMAX1 and SMXLs is still meager. It also remains to be elucidated by what mechanism the different SL stereoisomers exert different plant responses. Answering this question will require detailed investigations on the binding of the different SL stereoisomers by D14. Additional insights may be gained by the adaptation of *in vivo*-developed SL receptors and by screening for mutants in the SL-signaling pathway.

Acknowledgements

The author thanks Dr. Michael Melzer and Dr. Twan Rutten for critical reading of the text. The author is supported by scholarships funded by the Ministry of Science and Higher Education (424/STYP/10/2015 and DN/MOB/245/IV/2015).

Author details

Marek Marzec[1,2]

Address all correspondence to: marek.marzec@us.edu.pl

1 Department of Genetics, Faculty of Biology and Environmental Protection, University of Silesia, Katowice, Poland

2 Department of Physiology and Cell Biology, Leibniz Institute of Plant Genetics and Crop Plant Research (IPK), Stadt Seeland, Germany

References

[1] Cook CE, Whichard LP, Turner B, Wall ME, Egley GH. Germination of witchweed (*Striga lutea* Lour.): Isolation and properties of a potent stimulant. Science. 1966; **154**(3753):1189-1190

[2] Screpanti C, Yoneyama K, Bouwmeester HJ. Trigolactones and parasitic weed management 50 years after the discovery of the first natural strigolactone strigol: Status and outlook. Pest Management Science. 2016;**72**(11):2013-2015

[3] Akiyama K, Matsuzaki KI, Hayashi H. Plant sesquiterpenes induce hyphal branching in arbuscular mycorrhizal fungi. Nature. 2005;**435**(7043):824-827

[4] Soto MJ, Fernández-Aparicio M, Castellanos-Morales V, García-Garrido JM, Ocampo JA, Delgado MJ, et al. First indications for the involvement of strigolactones on nodule formation in alfalfa (*Medicago sativa*). Soil Biology and Biochemistry. 2010;**42**(2): 383-385

[5] Gomez-Roldan V, Fermas S, Brewer PB, Puech-Pagès V, Dun EA, Pillot JP, et al. Strigolactone inhibition of shoot branching. Nature. 2008;**455**(7210):189-194

[6] Umehara M, Hanada A, Yoshida S, Akiyama K, Arite T, Takeda-Kamiya N, et al. Inhibition of shoot branching by new terpenoid plant hormones. Nature. 2008;**455**(7210):195-200

[7] Waters MT, Gutjahr C, Bennett T, Nelson DC. Strigolactone signaling and evolution. Annual Review of Plant Biology. 2017;**68**(8):1-31. DOI: 10.1146/annurev-arplant-042916-040925

[8] Marzec M, Muszynska A, Gruszka D. The role of strigolactones in nutrient-stress responses in plants. International Journal of Molecular Sciences. 2013;**14**(5):9286-9304

[9] Ito S, Ito K, Abeta N, Takahashi R, Sasaki Y, Yajima S. Effects of strigolactone signaling on Arabidopsis growth under nitrogen deficient stress condition. Plant Signaling & Behavior. 2016;**11**(1):e1126031

[10] Bu Q, Lv T, Shen H, Luong P, Wang J, Wang Z, et al. Regulation of drought tolerance by the F-box protein MAX2 in Arabidopsis. Plant Physiology. 2014;**164**(1):424-439

[11] Van Ha C, Leyva-González MA, Osakabe Y, Tran UT, Nishiyama R, Watanabe Y, et al. Positive regulatory role of strigolactone in plant responses to drought and salt stress. Proceedings of the National Academy of Sciences. 2014;**111**(2):851-856

[12] Liu J, He H, Vitali M, Visentin I, Charnikhova T, Haider I, et al. Osmotic stress represses strigolactone biosynthesis in Lotus japonicus roots: Exploring the interaction between strigolactones and ABA under abiotic stress. Planta. 2015;**241**(6):1435-1451

[13] Ruiz-Lozano JM, Aroca R, Zamarreño ÁM, Molina S, Andreo-Jiménez B, Porcel R, et al. Arbuscular mycorrhizal symbiosis induces strigolactone biosynthesis under drought and improves drought tolerance in lettuce and tomato. Plant, Cell & Environment. 2016;**39**(2):441-452

[14] Visentin I, Vitali M, Ferrero M, Zhang Y, Ruyter-Spira C, Novák O, Cardinale F, et al. Low levels of strigolactones in roots as a component of the systemic signal of drought stress in tomato. New Phytologist. 2016;**212**(4):954-963

[15] Marzec M, Muszynska A. In silico analysis of the genes encoding proteins that are involved in the biosynthesis of the RMS/MAX/D pathway revealed new roles of strigolactones in plants. International Journal of Molecular Sciences. 2015;**16**(4):6757-6782

[16] Marzec M. Strigolactones as part of the plant defence system. Trends in Plant Science. 2016;**21**(11):900-903

[17] Al-Babili S, Bouwmeester HJ. Strigolactones, a novel carotenoid-derived plant hormone. Annual Review of Plant Biology. 2015;**66**:161-186

[18] Xie X, Yoneyama K, Kisugi T, Uchida K, Ito S, Akiyama K, et al. Confirming stereochemical structures of strigolactones produced by rice and tobacco. Molecular Plant. 2013;**6**(1):153-163

[19] Kretzschmar T, Kohlen W, Sasse J, Borghi L, Schlegel M, Bachelier JB, et al. A petunia ABC protein controls strigolactone-dependent symbiotic signalling and branching. Nature. 2012;**483**(7389):341-344

[20] Hou X, Rivers J, León P, McQuinn RP, Pogson BJ. Synthesis and function of apocarotenoid signals in plants. Trends in Plant Science. 2016;**21**(9):792-803

[21] Lin H, Wang R, Qian Q, Yan M, Meng X, Fu Z, et al. DWARF27, an iron-containing protein required for the biosynthesis of strigolactones, regulates rice tiller bud outgrowth. The Plant Cell. 2009;**21**(5):1512-1525

[22] Waters MT, Brewer PB, Bussell JD, Smith SM, Beveridge CA. The Arabidopsis ortholog of rice DWARF27 acts upstream of MAX1 in the control of plant development by strigolactones. Plant Physiology. 2012;**159**(3):1073-1085

[23] Alder A, Jamil M, Marzorati M, Bruno M, Vermathen M, Bigler P, et al. The path from β-carotene to carlactone, a strigolactone-like plant hormone. Science. 2012;**335**(6074):1348-1351

[24] Booker J, Auldridge M, Wills S, McCarty D, Klee H, Leyser O. MAX3/CCD7 is a carotenoid cleavage dioxygenase required for the synthesis of a novel plant signaling molecule. Current Biology. 2004;**14**(14):1232-1238

[25] Zou J, Zhang S, Zhang W, Li G, Chen Z, Zhai W, et al. The rice HIGH-TILLERING DWARF1 encoding an ortholog of Arabidopsis MAX3 is required for negative regulation of the outgrowth of axillary buds. The Plant Journal. 2006;**48**(5):687-698

[26] Sorefan K, Booker J, Haurogné K, Goussot M, Bainbridge K, Foo E, et al. MAX4 and RMS1 are orthologous dioxygenase-like genes that regulate shoot branching in Arabidopsis and pea. Genes & Development. 2003;**17**(12):1469-1474

[27] Arite T, Iwata H, Ohshima K, Maekawa M, Nakajima M, Kojima M, et al. DWARF10, an RMS1/MAX4/DAD1 ortholog, controls lateral bud outgrowth in rice. The Plant Journal. 2007;**51**(6):1019-1029

[28] Abe S, Sado A, Tanaka K, Kisugi T, Asami K, Ota S, et al. Carlactone is converted to carlactonoic acid by MAX1 in Arabidopsis and its methyl ester can directly interact with AtD14 in vitro. Proceedings of the National Academy of Sciences. 2014;**111**(50):18084-18089

[29] Zhang Y, Van Dijk AD, Scaffidi A, Flematti GR, Hofmann M, Charnikhova T, et al. Rice cytochrome P450 MAX1 homologs catalyze distinct steps in strigolactone biosynthesis. Nature Chemical Biology. 2014;**10**(12):1028-1033

[30] Brewer PB, Yoneyama K, Filardo F, Meyers E, Scaffidi A, Frickey T, et al. LATERAL BRANCHING OXIDOREDUCTASE acts in the final stages of strigolactone biosynthesis in Arabidopsis. Proceedings of the National Academy of Sciences. 2016;**113**(22):6301-6306

[31] Arite T, Umehara M, Ishikawa S, Hanada A, Maekawa M, Yamaguchi S, et al. D14, a strigolactone-insensitive mutant of rice, shows an accelerated outgrowth of tillers. Plant and Cell Physiology. 2009;**50**(8):1416-1424

[32] Waters MT, Nelson DC, Scaffidi A, Flematti GR, Sun YK, Dixon KW, et al. Specialisation within the DWARF14 protein family confers distinct responses to karrikins and strigolactones in Arabidopsis. Development. 2012;**139**(7):1285-1295

[33] Hamiaux C, Drummond RS, Janssen BJ, Ledger SE, Cooney JM, Newcomb RD, et al. DAD2 is an α/β hydrolase likely to be involved in the perception of the plant branching hormone, strigolactone. Current Biology. 2012;**22**(21):2032-2036

[34] Marzec M, Gruszka D, Tylec P, Szarejko I. Identification and functional analysis of the HvD14 gene involved in strigolactone signaling in Hordeum vulgare. Physiologia Plantarum. 2016;**158**(3):341-355

[35] Zheng K, Wang X, Weighill DA, Guo HB, Xie M, Yang Y, et al. Characterization of DWARF14 genes in Populus. Scientific Reports. 2016;**6**:21593

[36] Yao R, Ming Z, Yan L, Li S, Wang F, Ma S, et al. DWARF14 is a non-canonical hormone receptor for strigolactone. Nature. 2016;**536**:469-473

[37] Scaffidi A, Waters MT, Bond CS, Dixon KW, Smith SM, Ghisalberti EL, et al. Exploring the molecular mechanism of karrikins and strigolactones. Bioorganic & Medicinal Chemistry Letters. 2012;**22**(11):3743-3746

[38] Zhao LH, Zhou XE, Wu ZS, Yi W, Xu Y, Li S, et al. Crystal structures of two phytohormone signal-transducing α/β hydrolases: Karrikin-signaling KAI2 and strigolactone-signaling DWARF14. Cell Research. 2013;**23**(3):436-439

[39] Nakamura H, Xue YL, Miyakawa T, Hou F, Qin HM, Fukui K, et al. Molecular mechanism of strigolactone perception by DWARF14. Nature Communications. 2013;**4**:2613

[40] Zhao LH, Zhou XE, Yi W, Wu Z, Liu Y, Kang Y, et al. Destabilization of strigolactone receptor DWARF14 by binding of ligand and E3-ligase signaling effector DWARF3. Cell Research. 2015;**25**(11):1219-1236

[41] Chevalier F, Nieminen K, Sánchez-Ferrero JC, Rodríguez ML, Chagoyen M, Hardtke CS, et al. Strigolactone promotes degradation of DWARF14, an α/β hydrolase essential for strigolactone signaling in Arabidopsis. The Plant Cell. 2015;**26**(3):1134-1150

[42] Kameoka H, Dun EA, Lopez-Obando M, Brewer PB, de Saint Germain A, Rameau C, et al. Phloem transport of the receptor DWARF14 protein is required for full function of strigolactones. Plant Physiology. 2016;**172**(3):1844-1852

[43] Wang C, Liu Y, Li SS, Han GZ. Insights into the origin and evolution of the plant hormone signaling machinery. Plant Physiology. 2015;**167**(3):872-886

[44] Ishikawa S, Maekawa M, Arite T, Onishi K, Takamure I, Kyozuka J. Suppression of tiller bud activity in tillering dwarf mutants of rice. Plant and Cell Physiology. 2005;**46**(1):79-86

[45] Stirnberg P, Furner IJ, Ottoline Leyser HM. MAX2 participates in an SCF complex which acts locally at the node to suppress shoot branching. The Plant Journal. 2007;**50**(1):80-94

[46] Wang Y, Sun S, Zhu W, Jia K, Yang H, Wang X. Strigolactone/MAX2-induced degradation of brassinosteroid transcriptional effector BES1 regulates shoot branching. Developmental Cell. 2013;**27**(6):681-688

[47] Booker J, Sieberer T, Wright W, Williamson L, Willett B, Stirnberg P, et al. MAX1 encodes a cytochrome P450 family member that acts downstream of MAX3/4 to produce a carotenoid-derived branch-inhibiting hormone. Developmental Cell. 2005;**8**(3):443-449

[48] Zhao J, Wang T, Wang M, Liu Y, Yuan S, Gao Y, et al. DWARF3 participates in an SCF complex and associates with DWARF14 to suppress rice shoot branching. Plant and Cell Physiology. 2014;**55**(6):1096-1109

[49] Sheard LB, Tan X, Mao H, Withers J, Ben-Nissan G, Hinds TR, et al. Jasmonate perception by inositol-phosphate-potentiated COI1-JAZ co-receptor. Nature. 2010;**468**(7322):400-405

[50] Dharmasiri N, Dharmasiri S, Estelle M. The F-box protein TIR1 is an auxin receptor. Nature. 2005;**435**(7041):441-445

[51] Jiang L, Liu X, Xiong G, Liu H, Chen F, Wang L, et al. DWARF 53 acts as a repressor of strigolactone signalling in rice. Nature. 2013;**504**(7480):401-405

[52] Zhou F, Lin Q, Zhu L, Ren Y, Zhou K, Shabek N, et al. D14-SCFD3-dependent degradation of D53 regulates strigolactone signalling. Nature. 2013;**504**(7480):406-410

[53] Soundappan I, Bennett T, Morffy N, Liang Y, Stanga JP, Abbas A, et al. SMAX1-LIKE/D53 family members enable distinct MAX2-dependent responses to strigolactones and karrikins in Arabidopsis. The Plant Cell. 2015;**27**(11):3143-3159

[54] Wang L, Wang B, Jiang L, Liu X, Li X, Lu Z, et al. Strigolactone signaling in Arabidopsis regulates shoot development by targeting D53-like SMXL repressor proteins for ubiquitination and degradation. The Plant Cell. 2015;**27**(11):3128-3142

[55] Liang Y, Ward S, Li P, Bennett T, Leyser O. SMAX1-LIKE7 signals from the nucleus to regulate shoot development in Arabidopsis via partially EAR motif-independent mechanisms. The Plant Cell. 2016;**28**(7):1581-1601

[56] Ke J, Ma H, Gu X, Thelen A, Brunzelle JS, Li J, et al. Structural basis for recognition of diverse transcriptional repressors by the TOPLESS family of corepressors. Science Advances. 2015;**1**(6):e1500107

[57] Aguilar-Martínez JA, Poza-Carrión C, Cubas P. Arabidopsis BRANCHED1 acts as an integrator of branching signals within axillary buds. The Plant Cell. 2007;**19**(2):458-472

[58] Takeda T, Suwa Y, Suzuki M, Kitano H, Ueguchi-Tanaka M, Ashikari M, et al. The OsTB1 gene negatively regulates lateral branching in rice. The Plant Journal. 2003;**33**(3):513-520

[59] Doebley J, Stec A, Hubbard L. The evolution of apical dominance in maize. Nature. 1997;**386**(6624):485-488

[60] Braun N, de Saint Germain A, Pillot JP, Boutet-Mercey S, Dalmais M, Antoniadi I, et al. The pea TCP transcription factor PsBRC1 acts downstream of strigolactones to control shoot branching. Plant Physiology. 2012;**158**(1):225-238

[61] Minakuchi K, Kameoka H, Yasuno N, Umehara M, Luo L, Kobayashi K, et al. Fine Culm1 (FC1) works downstream of strigolactones to inhibit the outgrowth of axillary buds in rice. Plant and Cell Physiology. 2010;**51**(7):1127-1135

[62] Guan JC, Koch KE, Suzuki M, Wu S, Latshaw S, Petruff T, et al. Diverse roles of strigolactone signaling in maize architecture and the uncoupling of a branching-specific subnetwork. Plant Physiology. 2012;**160**(3):1303-1317

[63] Marzec M. Perception and signaling of strigolactones. Frontiers in Plant Science. 2016;**7**:1260

[64] Flematti GR, Dixon KW, Smith SM. What are karrikins and how were they 'discovered' by plants? BMC Biology. 2015;**13**(1):108

[65] Morffy N, Faure L, Nelson DC. Smoke and hormone mirrors: Action and evolution of karrikin and strigolactone signaling. Trends in Genetics. 2016;**32**(3):176-188

[66] Bythell-Douglas R, Waters MT, Scaffidi A, Flematti GR, Smith SM, Bond CS. The structure of the Karrikin-Insensitive Protein (KAI2) in *Arabidopsis thaliana*. PLoS One. 2013;**8**(1):e54758

[67] Guo Y, Zheng Z, La Clair JJ, Chory J, Noel JP. Smoke-derived karrikin perception by the α/β-hydrolase KAI2 from Arabidopsis. Proceedings of the National Academy of Sciences. 2013;**110**(20):8284-8289

[68] Waters MT, Scaffidi A, Sun YK, Flematti GR, Smith SM. The karrikin response system of Arabidopsis. The Plant Journal. 2014;**79**(4):623-631

[69] Waters MT, Scaffidi A, Moulin SL, Sun YK, Flematti GR, Smith SM. A Selaginella moellendorffii ortholog of Karrikin Insensitive2 functions in Arabidopsis development but cannot mediate responses to karrikins or strigolactones. The Plant Cell. 2015;**27**(7):1925-1944

[70] Waters MT, Scaffidi A, Flematti G, Smith SM. Substrate-induced degradation of the α/β-fold hydrolase Karrikin Insensitive2 requires a functional catalytic triad but is independent of MAX2. Molecular Plant. 2015;**8**(5):814-817

[71] Stanga JP, Smith SM, Briggs WR, Nelson DC. Suppressor of More Axillary Growth2 1 controls seed germination and seedling development in Arabidopsis. Plant Physiology. 2013;**163**(1):318-330

Chapter 6

Cross Talk between Nitric Oxide and Phytohormones Regulate Plant Development during Abiotic Stresses

Fahim Nawaz, Rana Nauman Shabbir,
Muhammad Shahbaz, Sadia Majeed,
Muhammad Raheel, Waseem Hassan and
Muhammad Amir Sohail

Additional information is available at the end of the chapter

http://dx.doi.org/10.5772/intechopen.69812

Abstract

Plants, being sessile, are concurrently exposed to various biotic and abiotic stresses. The perception of stress signals in plants involves a wide spectrum of signal transduction pathways that interact to induce tolerance against adverse environmental conditions. This functional overlapping among various stress signaling cascades also leads to the expression of genes that regulate biosynthesis or action of other hormones. Phytohormonal signals, activated by both developmental and environmental responses, play a crucial role to develop stress tolerance in plants. Nitric oxide (NO) is one of the major players in plant signaling networks. Emerging evidence supports that NO interplays with signaling pathways of auxins, gibberellins, abscisic acid, ethylene, jasmonic acid, brassinosteroids, and other plant hormones to control metabolism, growth, and development in plants. This chapter focuses on the current state of knowledge of cross talk between signaling pathways of NO and phytohormones in plants exposed to various abiotic stresses.

Keywords: nitric oxide, phytohormones, abiotic stresses, signaling cascades, plant growth

1. Introduction

Exposure to a wide array of environmental stresses is one of the most crucial factors that negatively influence plant growth and productivity worldwide. Plants respond to such adverse conditions through perception of endogenous and exogenous stress factors via hormone signaling networks along with the coordination of several downstream signal transduction mechanisms involving cyclic nucleotides, calcium ions, and reactive oxygen (such as hydrogen peroxide) or nitrogen (e.g., nitric oxide) species. Acclimation to abiotic stresses is achieved through turgor maintenance [1], accumulation of osmolytes [2], regulation of photosynthetic and transpiration rate, and activation of antioxidant machinery [3]. Moreover, stress-induced alterations in gene expression and metabolism stimulate several anti-stress compounds, which help to modify physiology, phenology, growth, and reproduction of plants exposed to adverse environmental conditions [4].

Nitric oxide (NO) is an important metabolite and stress signaling molecule that influences multitude of physiological and developmental functions in plants. It serves as a key component of the signaling cascades involved in plant growth, metabolism, and adaptive responses to various biotic and abiotic stresses. It is well established that NO regulates a plethora of physiological processes ranging from seed germination to plant senescence. Emerging evidence suggests this potential plant growth regulator interplays with various phytohormones (PHs) to control metabolism, growth, and development in plants.

During the last few years, extensive research has been carried out to explore the multiple and diversified mechanisms underlying PHs interactions with NO. There is virtually no doubt that NO acts either upstream or downstream of PHs [5, 6]. It seems that NO modulates the biosynthesis, distribution, degradation, and conjugation of elements involved in PHs transport and signaling [7–11]. However, further studies are required to explain how NO concomitantly interacts with hormone-related proteins at post-transcriptional or even translational level. Similarly, the understanding of mechanisms underlying intersection of NO signaling with signaling cascades of auxins (AUXs), gibberellins (GBs), cytokinins (CKs), ethylene (ETs), absicic acid (ABA), salicylic acid (SA), jasmonic acid (JA), polyamines (PAs), brassinosteroids (BRs), and strigolactones (SLs) under abiotic stress conditions remains elusive. Considering the common function played by these plant growth regulations in enhancing plant tolerance to biotic and abiotic stresses, it can be speculated that PHs-mediated stress responses are linked with NO synthesis. Therefore, this chapter would focus on the current state of knowledge of cross talk between signaling pathways of NO and PHs in plants exposed to various abiotic stresses (**Table 1**).

Type of stress	Phytohormone	Plant species	Response	Relation with NO	References
Drought stress	ABA	*Zea mays*	Increased expression of ABA biosynthetic gene *vp14*	+	Zhang et al. [26]
	AUX	*Tagetes erecta*	Development of adventitious roots	+	Liao et al. [33]
	SA	*Triticum aestivum*	Increased tolerance against osmotic stress	+	Alavi et al. [41]
	CK	*Zea mays*	Regulation of photosynthetic machinery	+	Shao et al. [42]

http://dx.doi.org/10.5772/intechopen.69812

Type of stress	Phytohormone	Plant species	Response	Relation with NO	References
Cd toxicity	ET	*Pisum sativum*	Promoted the Cd-induced senescence processes	–	Rodríguez-serrano et al. [64]
	PAs	*Triticum aestivum*	Inhibition of root growth	+	Groppa et al. [66]
	AUX	*Arabidopsis thaliana*	Stabilization of AUX repressor protein IAA17 through suppression of AUX carriers PIN1/3/7	–	Yuan and Huang [67]
		Medicago truncatula	Improved antioxidative capacity and reduced degradation of AUX in roots	+	Xu et al. [68]
	SA	*Lolium perenne*	Increased activities of antioxidative enzymes	+	Wang et al. [72]
		Arachis hypogaea	Restricted Cd distribution to organelles	+	Xu et al. [78]
Ni toxicity	SA	*Brassica napus*	Enhanced chlorophyll contents and reduced lipid peroxidation and proline accumulation	+	Kazemi et al. [77]
Cu toxicity	BR	*Raphanus sativus*	Increased ABA synthesis resulted in improved tolerance	+	Choudhary et al. [80]
Al toxicity	GA	*Triticum aestivum*	Promoted apical root growth	+	He et al. [50]
Salinity stress	ABA	*Gossypium hirsutum*	Decreased salt-induced leaf senescence by regulating the expression of ABA biosynthesis genes (*NCED2 and NCED9*)	–	Kong et al. [88]
	ET	*Lycopersicon esculentum*	Reduced ROS levels and blocked ET synthesis resulting in lower dead cell ratio in cell suspension cultures	–	Poór and Tari [85]
	AUX	*Arabidopsis thaliana*	Repressed AUX signaling through stabilization of AUXIN RESISTANT3 (AXR3)/INDOLE-3-ACETIC ACID17 (IAA17)	–	Liu et al. [94]
	SA	*Fagus sylvatica*	Reduced H_2O_2 accumulation, limited Na^{2+} uptake and increased influx of H^+-ATPase to plasma membrane	+	Dong et al.[97]
	PAs	*Cucumis sativus*	Reduced free putrescine, spermidine and polyamine oxidase (PAO) activity	–	Fan et al. [103]

Type of stress	Phytohormone	Plant species	Response	Relation with NO	References
Temperature stress	ABA	*Phragmites australis*	Improved the thermotolerance of plant calluses	+	Song et al. [109]
		Medicago sativa	Enhanced *MfSAMS1* expression to increase acclimation against cold stress	+	Guo et al. [123]
	PAs	*Lycopersicon esculentum*	Increased putrescine and spermidine levels and stimulated the expression of genes encoding Spd synthase (*LeSPDS*), arginine decarboxylase (*LeADC. LeADC1*) and ornithine decarboxylase (*LeODC*) to improve chilling stress tolerance	+	Diao et al. [121]
		Zingiber officinale	Conversion of putrescine into spermidine or spermine conferred cold tolerance	+	Li et al. [124]
	SA	*Spinacia oleracea*	Increased NR activity reduced chilling injury	+	Aydin and Nalbantoğlu [128]
	JA	*Cucumis sativus*	Increased CAT activity to scavenge H_2O_2, leading to reduced chilling injury	+	Liu et al. [129]

Table 1. Summary of representative reports on the interaction of nitric oxide with phytohormones during various abiotic stresses.

2. NO-phytohormone cross talk under drought stress

Drought stress is one of the major limiting factors affecting multiple aspects of plant growth and productivity [2]. The typical mechanism of plants response to water stress, frequently caused by drought, is closure of stomata to conserve water. NO and ABA are the two most important stress-related molecules that intensively cross talk during environmental challenges like drought to induce plant adaptive responses such as stomatal closure and activation of antioxidant machinery [5, 11]. Evidence suggests that NO acts downstream of ABA as decreased NO synthesis reduces ABA-induced responses in plant tissues exposed to stress conditions [12, 13]. However, NO is also reported to counteract ABA during events not linked to stress adaptation such as breaking of seed dormancy [14, 15]. It indicates a certain level of specificity in NO-ABA cross talk mechanisms, which seems to depend on the type of plant cell, tissue or organ studied, or nature of physiological event under analysis.

Generation of ROS (H_2O_2) under adverse environmental conditions triggers NO-mediated ABA responses such induction of stomatal closure [16], activation of antioxidant enzymes [17], and up-regulation of transcription factors [18]. In addition, cGMP and type 2C protein

http://dx.doi.org/10.5772/intechopen.69812

phosphatases (PP2Cs) have also been identified to participate in downstream of NO-mediated ABA signal transduction and upstream of cytosolic Ca^{2+} during the regulation of stomatal apparatus [19–21]. Moreover, the calcium/calmodulin system and mitogen-activated protein kinases (MAPKs) have also been demonstrated as key downstream elements involved in ABA or H_2O_2-induced NO signaling during plant antioxidant defense mechanisms [22, 23]. Cross talk between NO and ABA in the ABA-dependant signaling network up-regulated the cytosolic Ca^{2+} to regulate Crassulacean acid metabolism (CAM) expression in bromeliads that significantly improved plant tolerance in a water-limited environment [21, 24]. It seems that ABA-induced NO production is associated with increased nitrate reductase (NR) activity that controls stomatal movements in *Arabidopsis* [19] and CAM expression in bromeliads [24]. The expression of CYP707A2 gene, induced by NO biosynthesis, during seed germination initiated ABA catabolism and increased ABA levels to enhance plant resistance against drought stress [11, 25]. BR treatment of water-stressed *Zea mays* leaves induced NO generation in mesophyll cells and up-regulated the expression of ABA biosynthetic gene *vp14* to enhance water stress tolerance in *Zea mays* [26].

Interestingly, NO serves as a second messenger in the signaling cascades of various plant hormones such as GA, JA, ET, CK, and AUX involved in the regulation of stomata under environmental stress conditions [27, 28]. Interactions between NO and AUX signaling pathways are complex and need to be explored in plants exposed to water-limited environment. It is well established that both NO and AUX interplay during growth and development of plant roots [29, 30]. Association of AUX with ET to regulate root morphology and development is considered a key aspect of drought tolerance in plants [31]. Development of adventitious roots in cucumber hypocotyl cuttings involves the cross talk between AUX and NO signaling networks activated by Ca^{2+} dependent protein kinase activity [32]. Since NO is intensively involved in lateral root formation during drought stress [33], it may be speculated that AUX and NO signaling cascades interact and influence the architecture and development of root hair and root meristem size [34, 35] for the extraction of more water under drought stress conditions.

Drought stress influences the signaling of various JA-associated genes [36]. JA stimulates CDPK production by increasing Ca^{2+} influx and the resultant signal cascade results in ABA-regulated stomatal closure. A rapid loss in turgor and subsequent reduction in stomatal aperture were noted in excised *Arabidopsis* leaves treated with either ABA or methyl JA (MeJA) [37]. Suppression of MeJA-induced Ca^{2+} oscillations in guard cells of ABA-deficient mutants [38] implies that MeJA cross talk with ABA involves Ca^{2+} signal transduction pathways. Moreover, treatment with ABA or MeJA induces the formation of NO and ROS in guard cells [37]. Studies involving *Arabidopsis* revealed that ABA mediated Ca^{2+} influx into cytoplasm involves CPK6, which acts downstream of NO and ROS signaling and therefore may be a target of NO-stimulated Ca^{2+} influx into the cytoplasm [39]. In 2008, Palmieri et al. demonstrated that NO treatment up-regulated several genes involved in the JA biosynthetic pathway, indicating a potential regulation of JA signaling through the ROS/NO pathway [40]. NO also acts downstream of SA signaling to scavenge ROS in water-stressed plants. Coordinated action of NO and SA was found to alleviate the damaging effects of polyethylene (PEG)-induced osmotic stress in *Triticum aestivum* seedlings [41].

A positive interaction between NO and CK under water-limited environment was reported by Shao et al. [42]. Treatment of plants with CK plus NO scavenger (Hemoglobin) revealed that CK promoted NO signaling, probably mainly through a NR source in plants exposed to water stress conditions. CK interaction with NO signaling cascades regulated photosynthetic machinery and increased the adaptability to drought stress in *Zea mays* [42]. Contrasting reports indicate antagonistic interaction between NO and CK, for example, CK-induced reduction in NO levels promoted stomatal opening in dark grown *Vicia fabia* seedlings [43]. Presumably, CKs activate plasma membrane H1-ATPase through decreasing NO levels in guard cells, and then stimulate stomatal opening in darkness. Wilhelmova et al. [44] reported similar results in transgenic tobacco plants. Negative interaction between NO and CKs was evident during leaf development as increased NO production reduced CKs level in aging leaves. Evidence suggests that reaction of zeatin with peroxynitrite, a NO derivative, reduces its availability in plants [45].

3. NO-phytohormone cross talk under heavy metals stress

Heavy metals (HMs) are phytotoxic elements that can damage plant growth and metabolism at very low concentrations [46]. The involvement of plant hormones such as IAA, CK, and ET to alleviate HMs-induced toxicity is well reported [47–49]. Some recent studies suggest that NO acts in concert with signaling pathways of phytohormones to induce tolerance against excess elements [50, 51]. However, the exact nature of NO-hormone interactions still needs to be explored and is largely dependent on the species, the plant organ as well as concentration of metal and duration of stress [52].

Cadmium is one of the most widely distributed HM in agricultural soils [53]. Cd-induced increase in endogenous levels of NO is associated with its role as a bioactive molecule to quench ROS [54]. Alterations in hormonal homeostasis are potential signals that directly affect plant responses to Cd stress, including interplay between hormones and the whole plant signaling network, such as the ROS [55], MAPK [56], and NO signaling pathways [57]. Exposure to short-term Cd stress revealed an interrelation of ET with NO generation, polyamine metabolism, and MAPK cascades in young *Glycine max* seedlings [58]. It is well documented that exposure to HMs enhances the production of ET [59] due to increased 1-aminocyclopropane-1-carboxylic acid (ACC) synthase (ACS) activity in metal stressed plants [60]. Cross talk between ET and stress signaling molecules like NO is important to understand the mechanisms of plant adaptation to HM-induced oxidative stress [61–63]. More recently, Thao et al. [51] suggested a possible link between NO and ET through MAPKs in plants exposed to HM stress. Accumulation of ET reduced NO levels and promoted the Cd-induced senescence processes in *Pisum sativum* [64]. Similarly, integration of ET, NO, PA, and MAPKs pathways improved tolerance in young *Glycine max* seedlings against short-term Cd stress [52]. The potential involvement of ETHYLENE INSENSITIVE2 (EIN2) in improving resistance against lead (Pb) stress has also been demonstrated in *Arabidopsis* [65]. Treatment of *Triticum aestivum* seedlings with Cd and PAs (spermine and putrescine) was found to induce NO generation in roots leading to root growth inhibition [66].

Experiments with *Arabidopsis* showed that Cd toxicity triggers NO accumulation, which in turn promotes the stabilization of AUX repressor protein IAA17 through suppression of AUX carriers PIN1/3/7 [67]. Xu et al. [68] found that exogenous NO supply improved the antioxidative capacity and reduced the degradation of AUX in roots of *Medicago truncatula* seedlings exposed to Cd stress. In another study, it was noted that NO acts downstream of AUX on modulating root architecture in *Arabidopsis seedlings* exposed to Cd stress [69]. Rodríguez-Serrano et al. [70] demonstrated that Cd toxicity strongly depressed the activity of NO-synthase dependent NO production in *Pisum sativum* seedlings; however, exogenous Ca supply ameliorated this effect due to enhanced JA and ET production. Interplay among NO, H_2O_2, and SA has also been reported in lupine seedlings [71] and ryegrass plants [72] under Cd stress. Application of putrescine (Put) and NO in combination was reported to alleviate Cd toxicity in *Vigna radiate* by triggering the activity of both enzymatic and nonenzymatic antioxidant machinery and a parallel increase in phytochelatin synthesis [73].

Interplay between NO and GA has been reported to influence a wide spectrum of physiological processes, including seed germination, primary root growth, and inhibition of hypocotyl elongation [8, 29]. Interaction of NO with GA was observed to promote apical root growth in *Triticum aestivum* roots exposed to aluminum (Al) toxicity [50]. Contrary reports of Zhu et al. [74] showed that GA mediated alleviation of Cd toxicity in *Arabidopsis* was linked to reduction of Cd-induced NO accumulation and suppression of up-regulation of *IRT1*. Antagonistic relationship between NO and GA was also reported by Wu et al. [75] who noticed that NO accumulation inhibited the stimulatory effect of GA on primary root growth under low phosphorous (P) conditions. They suggested that GA interacts with NO and P pathways on DELLA-SLY module. Studies have shown the positive interaction between NO and CK, for example, Shen et al. [76] found that NO deficiency inhibited the activation CK-induced gene CYCLIN-D3;1 (CYCD3;1), which in turn promoted callus initiation from somatic plant tissues. Recent reports of Liu et al. [45] revealed that CK directly interact with NO to reduce endogenous NO levels in *Arabidopsis* implying the protective role of CK against nitrosative stress; however, no study has yet provided definitive evidence for NO and CK interaction under metal stress.

Combined NO and SA application was observed to counteract the toxic effects of Ni in *Brassica napus* through enhanced chlorophyll contents and reduced lipid peroxidation and proline accumulation [77]. Similarly, NO and SA increased Cd contents in cell walls of Cd-stressed *Arachis hypogaea* that reduced the distribution of Cd to organelles [78]. In recent years, accumulating evidence indicates the involvement of BRs induced NO production in root architecture and development [79]. Choudhary et al. [80] found that BR induced increased NO production promoted ABA synthesis that alleviated the toxic effects of Cu in *Raphanus sativus* seedlings.

Cross talk between plant hormones and NO is also considered critical for Fe-deficiency signaling [81]. Evidence obtained in *Arabidopsis* suggests interplay between ET and NO for up-regulation of genes (A*tFIT, AtbHLH39, AtFRO2, AtNAS1, AtNAS2, AtFRD3, AtMYB72*) related to Fe-deficiency [82]. Although, it is well reported that enhanced NO generation helps to maintain root growth under Cd stress [83]; however, the exact role of NO under excess Fe conditions is rudimentary and demands further investigation.

4. NO-phytohormone cross talk under salinity stress

Salinity stress is considered one of the most harmful stresses due to its high magnitude and worldwide distribution [84]. Phytohormones play a key role in enhancing the tolerance and adaptability of plants against salinity stress. Some recent studies suggest that NO acts in concert with signaling pathways of phytohormones to induce tolerance against salt stress [85, 86]. Presumably, plant hormones such as ABA, ET, and AUX are transported from salt-treated roots to leaves to trigger NO synthesis or transport throughout the plant [87]. NO-induced alleviation of oxidative damage in salt-stressed plants is associated with increased antioxidant activities and decreased thiobarbituric acid reactive substances content [69]. ABA stimulates H_2O_2 accumulation that results in increased NO generation, leading to the activation of MAPK and up-regulation of genes associated with antioxidant enzymes [17, 18] in plants exposed to abiotic stresses like salinity. However, NO does not always positively interplay with ABA. In cotton, exogenous NO supply (using SNP as NO donor) reduced salt-induced leaf senescence by decreasing ABA content and down regulating the expression of ABA biosynthesis genes (*NCED2* and *NCED9*) [88].

In general, it is believed that ET biosynthesis corresponds to increased damage in plants. However, recent studies indicate ET as a stress-signaling hormone that interacts with signaling cascades of other phytohormones to enhance tolerance against various biotic/abiotic stresses [70, 86]. Studies involving tobacco seedlings showed that transcriptional activation of ethylene response factor (ERF) in ethylene-signaling process improved salt stress tolerance by decreasing ROS accumulation [89]. Treatment of *Arabidopsis* callus with 100 mM NaCl triggered the accumulation of NO that promoted ET emission, resulting in increased expression of the plasma membrane H^+-ATPase genes [90]. Hence, both NO and ET participate in up-regulation of plasma membrane H^+-ATPase that modulates ion homeostasis for improved salt tolerance. NO and ET also cooperate to stimulate the alternative respiratory pathway under salt stress conditions [91, 92]. Contrary reports of Poór and Tari [85] showed antagonistic relationship between NO and ET in tomato cell suspension cultures treated with NaCl (100 and 250 mM). Increased ET synthesis promoted ROS generation leading to high dead cell ratio in salt-stressed cell culture. However, NO generation decreased ROS levels and blocked ET synthesis resulting in lower dead cell ratio. In another study, absence of ET and NO in apical root segments and cell suspension culture, respectively, caused ionic imbalance (Na^+/K^+) that resulted in increased susceptibility to salinity stress [93]. AUX and NO involvement in inhibition of root meristem growth in salt-stressed *Arabidopsis* was revealed by Liu et al. [94]. They reported that salinity stress repressed root meristem growth by inhibiting the expression of *PINFORMED* (*PIN*) genes, thereby reducing AUX levels. Moreover, stabilization of AUXIN RESISTANT3 (AXR3)/INDOLE-3-ACETIC ACID17 (IAA17) repressed AUX signaling via NO accumulation.

Participation of both NO and ROS in SA-induced stomatal closure is also reported in literature [95]. Activation of a peroxidase (sensitive to the inhibitor salicylhydroxamic acid) by SA promotes ROS accumulation and NO generation in guard cells, leading to stomatal closure. Experiment with soybean seedlings showed that combined application of SNP (as NO donor) and SA alleviated the toxicity of NaCl-induced salt stress by increased proline accumulation and activation of CAT, APX, and GPX. Similar results were reported by Liu et al. [96] and Dong et al. [97] in *Gossypium hirsutum* and *Fagus sylvatica*, respectively. SA interaction with signaling cascades of NO modulated photosynthetic machinery and reduced H_2O_2 accumulation that promoted the influx of H^+-ATPase to plasma membrane. Moreover, synergistic effect of SA and NO improved Ca^{2+}/Mg^{2+} absorption and reduced Na^{2+} uptake under salt stress conditions [97].

Sulfur (S) is a major component of metabolites such as reduced glutathione (GSH), coenzyme A, methionine, cysteine (Cys), sulfo-lipids, iron-sulfur (Fe-S) clusters, and thioredoxin system involved in regulation of physiological processes under salt stress conditions [98]. Evidence suggests that NO promotes S-assimilation, which is linked to ET production through Cys synthesis [86]. Hence, it may be speculated that NO and S interact to modulate ABA and ET levels in guard cells that may influence the stomatal and photosynthetic response under salt stress conditions. NO combines with GSH to generate S-nitrosoglutathione (GSNO), leading to enhanced S requirement of plants for improved tolerance under environmental stress conditions [99, 100]. Coordinated effect of NO and S regulated the utilization of S and GSH resulting in improved growth and photosynthetic activity in salt-stressed mustard plants [86]. NO is a key regulatory signal that activates several biochemical processes and interacts with sulfhydryl groups and nitro groups in the process of nitration to enhance tolerance against salt stress [101]. NO also cooperates with other signaling molecules such as H_2S to enhance tolerance against salinity stress in plants. NO and H_2S cross talk helped to maintain low Na^+ levels with up-regulation of *HvHA1* and *HvSOS1* and increased plasma membrane H^+-ATPase levels in cytoplast of salt stressed barley seedlings [102]. Moreover, transcriptional activation of vacuolar Na^+/H^+ antiporter (*HvVNHX2*) and H^+-ATPase subunit β (*HvVHA-β*) up-regulated the expression of vacuolar Na^+/H^+ antiporter (NHE1) that helped to modulate Na^+ compartmentation into the vacuoles.

Recently, it has been proposed that NO negatively regulates CK signaling by limiting phosphorelay activity via S-nitrosylation [103]. Contrasting reports of Kong et al. [88] showed that foliar applied SNP (as NO donor) delayed salt-induced leaf senescence in cotton seedlings by up-regulating the expression of CK biosynthesis gene, isopentenyl transferase (*IPT*). NO induced reduction in total free PAs, free Put, spermidine (Spd), and polyamine oxidase (PAO) activity that was reported by Fan et al. [103] in cucumber seedlings exposed to NaCl stress. These reports provide a strong evidence for NO cross talk with plant hormones to induce salt stress tolerance in plants; however, further in depth studies to understand interplay among these plant growth regulators in terms of transcriptional regulation, signal transduction, and ion detoxification are needed.

5. NO-phytohormone cross talk under temperature stress

Temperature stress negatively influences the vegetative and reproductive growth phases of plants. Coordinated action between NO and plant hormones (ABA, JA, GA, CK) induce thermotolerance in plants by activating the antioxidant machinery and up-regulating the expression of genes encoding heat shock proteins [104–106]. Studies involving *Arabidopsis* mutants impaired in ABA biosynthesis (*aba1-1*) and signaling (*abi 1-1*) showed that drought and heat stress induced stomatal closure involved JA and H_2O_2 signaling that triggered NO levels [106] and Ca^{2+} and SLAC1 function [107, 108]. However, SA antagonized JA function to induce stomatal opening in *abi1-1* [106]. In *Phragmites communis*, ABA treatment triggered NOS activity and increased NO levels that improved the thermotolerance of plant calluses [109]. Treatment of *Stylosanthes guianensis* seedlings with ABA stimulated the activities of CAT, SOD, and APX suggesting that ABA-induced NO generation leads to the production of antioxidant enzymes [110]. Evidence supports the antagonist relationship between SA and ET in improving heat tolerance in plants by increasing proline contents and enhancing photosynthetic-NUE [111]. SA cross talk with AUX, ET, JA, and BR has been demonstrated in specific bioassays [112]. SA triggered increase in GST activity was noted to induce heat stress tolerance in *Zea mays* [113]. Presumably, SA reduced H_2O_2 accumulation through NO generation; however, direct evidences of NO interaction with plant hormones (SA, GA, AUX, BR, and JA) in improving plant heat stress tolerance are lacking. BRs are also thought to interact with ABA, SA, and ET to induce heat stress signaling through complex networks [114, 115]. BR treatment of *Brassica napus* seedlings subjected to short-term heat shocks was noted to enhance endogenous ABA concentration [116]. BR induced increase in ABA level has also been reported in cellular culture of *Chlorella vulgaris* [117].

Low temperature severely restricts plant growth and causes both structural and metabolic damages in plants [118]. Exposure to low temperature induces oxidative and nitrosative stress thereby promoting NO synthesis [119], which serves as a potential link between PA and ABA to induce stress responses in plants [120]. Literature indicated extensive cross talk among NO, ABA, PAs, and H_2O_2 to modulate various physiological and stress responses under low temperature conditions [110, 121]. Interplay among NO, SA, and ABA was noted to enhance the antioxidative activities (CAT, SOD, POX) that contributed to improved chilling injury in *Zea mays* seedlings [122]. Guo et al. [123] found that coordinated action between NO and ABA up-regulated cold-induced *MfSAMS1* expression, resulting in enhanced acclimation against cold stress in *Medicago sativa* subsp. *falcata*. Moreover, expression of *MfSAMS1* altered the levels of Spm, Put, and Spd and activities of PAO and copper-containing amine oxidase, which regulate anti-oxidant machinery during cold acclimation. Exogenous NO supply increased Put and Spd levels and stimulated the expression of genes encoding Spd synthase (*LeSPDS*), arginine decarboxylase (*LeADC. LeADC1*), and ornithine decarboxylase (*LeODC*) to improve chilling stress tolerance in *Lycopersicon esculentum* leaves. However, the expression of genes encoding Spm synthase (*LeSPMS*) and *S*-adenosylmethionine decarboxylase (*LeSAMDC*) was not influenced by NO treatment [121]. Reports of Li et al. [124] showed that NO treatment converts Put into Spd or Spm to confer cold tolerance in *Zingiber officinale* seedlings. Pretreatment of *Orzya sativa* seedlings with various ammonium concentrations decreased the effects of cold stress by increasing Put and Spd contents [125], suggesting the possible involvement of NO in stress tolerance. In a recent article, Wang et al. [126] reported the coordinated action of NO and PAs to induce chilling tolerance in cold-stored banana. NO treatment increased the activities of PAO, diamine oxidase (DAO) and glutamate decarboxylase (GAD), leading to γ-aminobutyric acid (GABA) accumulation to prevent chilling injury in fruits.

NR and NOS pathway are the most widely known NO sources in plants [19, 127]. Evidence obtained by Aydin and Nalbantoğlu [128] showed that SA pretreatment of *Spinacia oleracea* leaves influenced NR activity to induce chilling stress tolerance. A recent study indicated the involvement of JA in NO synthesis that increased CAT activity to scavenge H_2O_2, leading to reduced chilling injury in *Cucumis sativus* [129]. Therefore, it is concluded that NO cross talk with other hormones safeguards the quality of stored fruits and vegetables. Another study on NO revealed that it increases the expression of *MaCAT, MaPOD, MaSOD,* and *MaAPX* genes to alleviate damages caused by low temperature in banana (Wu et al. [75]). In *Elymus nutans,* interaction between NO and 5-aminolevulinic acid (ALA) stimulated antioxidant defense to reduce chilling injury [130]. Further investigations involving influence of NO on BR, CK, JA, and ET pathways are suggested which would provide important information about signaling cascades of these regulatory substances in cold stressed plants.

6. NO-phytohormone cross talk under other abiotic stresses

Ever increasing human population and industrial productivity has resulted in alarming rise in air pollutants, causing extensive damages to natural habitats of plant [131]. Ozone is characterized as one of the most phytotoxic air pollutants severely restricting plant growth and development [132]. Plants use many transportable chemical signals such as NO to turn the sensing of ozone from guard cells to adjacent epidermal and mesophyll cells [133]. Presumably, NO generation in relation to ozone stress induces ET and ABA synthesis and interferes with stomatal ABA response, potentially by inhibiting K^+ efflux at the guard cells [134]. The involvement of alternative oxidase (*AOX*) in the inhibition of ozone-induced toxicity has also been demonstrated to require both NO- and ET-dependent pathways [135]. Interestingly, Rao and Davies [136] observed that NO treatment caused leaf injury due to increased levels of ozone-induced ET production. Both SNP and ozone treatment up-regulated the expression of the ET biosynthesis related genes (*ACS6* and *ACC oxidase*), which correlates with ET formation [137]. In *Arabidopsis,* exogenous NO supply in combination with ozone stress was noted to attenuate the induction of SA biosynthesis and other defense-related genes [132].

Destruction of ozone layer in upper atmosphere, as a result of increased concentrations of air pollutants, has exposed living organisms to UV-radiation particularly UV-B that induces oxidative stress in plants [138, 139]. Although it is well known that NO interacts with ABA, ET, MeJA to control guard cell signaling in response to various environmental stresses [140, 141], only few reports are available with regard to NO, ET, and ABA cross talk in stomatal regulation under UV-B stress [142]. Studies involving *Lactuca sativa* seedlings showed that exogenous NO supply (using SNP as a NO donor) prevented UV-B induced inhibition of GA and IAA synthesis [143]. NO stimulated decrease in SA and ABA levels was found to be associated with reduced H_2O_2 and malondialdehyde contents. In contrast, coordinated action of NO and SA was observed to reduce UV-B stress in *Triticum aestivum* seedlings [144].

A transient NO burst is among the earliest responses to wounding [145]. NO production in wounded parts involves several pathways including cross talk with signaling cascades of hormones and endogenous signals [146, 147]. It was shown that NO and AUX actively take part in wound-healing response in plants [145, 148]. Imanishi et al. [149] presented evidence for the involvement MeJA and mechanical wounding in expression of the *Ipomoelin* gene (*IPO*) in sweet potato. Later, Jih et al. [150] demonstrated that SNP-derived NO delayed wounding-induced *IPO* expression, providing evidence for antagonistic association between NO and JA. In *Arabidopsis*, NO treatments led to elevated expression of key enzymes of the octadecanoid pathway, like *LOX2, AOS, or OPR3,* in wounded leaf epidermis [151]. However, this induction did not influence JA responsible genes, like PDF*1.2,* hence supporting the earlier evidences about NO and JA association. NO-induced wound-responses could act as a modulator of cell death initiation together with H_2O_2 accumulation, and delay of IPO-expression [152]. Contrasting reports in *Lycopersicon esculentum* demonstrated neither wound-induced NO burst, nor NO-induced elevation of endogenous SA levels [153]. Moreover, SNP-derived NO inhibited the expression of the proteinase inhibitors *Inh1, Inh2*, cathepsin D inhibitor (*CDI*), and metallocarboxypeptidase inhibitor (*CPI*) and increased *AOS* or *LOX* activity. Nevertheless, these studies demonstrate clearly that induction of a wound-response in plants involve cross talk among various stress signaling molecules.

Initiation of senescence in plants is controlled by various factors such as nutrient supply, light conditions, leaf age, and environmental stress [154]. Plant hormones such as ET and CK influence senescence by either promoting or delaying the process, respectively [155, 156]. Evidence supports the interaction of NO with other plant hormones to floral senescence and fruit maturation [157]. Recently, Ji et al. [158] demonstrated that SA treatment at low concentrations induced NOA1-dependent NO signaling and activated antioxidant defense to counteract MeJA-induced leaf senescence. NO plays a conceivable role to counteract the ABA- and jasmonate-induced senescence in rice by inhibiting H_2O_2 accumulation and lipid peroxidation [159]. Mishina et al. [160] found that delayed leaf senescence in *Arabidopsis* involves NO-induced reduction in SA levels. During fruit ripening, NO cross talk with SA and ET involves the regulation of levels of secondary metabolites such as anthocyanins [161]. NO-induced suppression of cell wall softening related enzymes such as polygalacturonase (PG), pectin methylesterase (PME), and pectate lyase (PL) was found to delay softening and ripening of stored *Carica papaya* by reducing ABA, IAA and zeatin ribose (ZR) levels [123].

7. Conclusion and future perspectives

Although our understanding of NO interactions with plant hormones has increased dramatically in past few years, many pieces of the puzzle are still missing. It is well established that NO coordinates with plant hormones to regulate gene expression and activities of anti-oxidative enzymes under adverse environmental conditions. However, our current knowledge about NO-phytohormone interactions is derived chiefly from NO-induced posttranslational modifications of transcription factors and biosynthetic enzymes. Future work is needed to explore the interplay among NO, plant hormones, ROS, protein kinases, and cytoskeletal proteins in order to understand the complicated network of NO signaling under abiotic stress conditions. Interestingly, most of the studies related to NO-phytohormonal interactions involve experiments in controlled laboratory environments, very little is known about the cross talk between these signaling molecules during flower initiation or grain development. Moreover, plants growing under natural conditions face multiple stresses; hence, future studies will need to address how NO interacts with the signaling cascades of phytohormones in plants exposed to two or more abiotic stresses.

Author details

Fahim Nawaz[1]*, Rana Nauman Shabbir[2], Muhammad Shahbaz[3], Sadia Majeed[4], Muhammad Raheel[5], Waseem Hassan[6] and Muhammad Amir Sohail[1]

*Address all correspondence to: fahim5382@gmail.com

1 Department of Agronomy, MNS University of Agriculture, Multan, Pakistan

2 Department of Agronomy, Agriculture College, Bahauddin Zakariya University, Multan, Pakistan

3 Department of Food Science and Technology, MNS University of Agriculture, Multan, Pakistan

4 Department of Agronomy, UCA & ES, The Islamia University of Bahawalpur, Pakistan

5 Department of Plant Pathology, UCA & ES, The Islamia University of Bahawalpur, Pakistan

6 Department of Soil Science, MNS University of Agriculture, Multan, Pakistan

References

[1] Waraich EA, Ahmad R, Ashraf MY. Role of mineral nutrition in alleviation of drought stress in plants. Australian Journal of Crop Science. 2011;**5**:764-777

[2] Shabbir RN, Waraich EA, Ali H, Nawaz F, Ashraf MY, Ahmad R, Awan MI, Ahmad S, Irfan M, Hussain S, Ahmad Z. Supplemental exogenous NPK application alters biochemical processes to improve yield and drought tolerance in wheat (*Triticum aestivum* L.). Environmental Science and Pollution Research. 2016;**23**:2651-2662

[3] Nawaz F, Ashraf MY, Ahmad R, Waraich EA, Shabbir RN, Bukhari MA. Supplemental selenium improves wheat grain yield and quality through alterations in biochemical processes under normal and water deficit conditions. Food Chemistry. 2015;**175**:350-357

[4] Penuelas J, Sardans J, Estiarte M, Ogaya R, Carnicer J, Coll M, Barbeta A, Rivas-Ubach A, Llusia J, Garbulsky M, Filella I. Evidence of current impact of climate change on life: A walk from genes to the biosphere. Global Change Biology. 2013;**19**:2303-2338

[5] Hancock JT, Neill SJ, Wilson ID. Nitric oxide and ABA in the control of plant function. Plant Science 2011;**181**:555-559

[6] Simontacchi M, García-Mata C, Bartoli CG, Santa-María GE, Lamattina L. Nitric oxide as a key component in hormone-regulated processes. Plant Cell Reports. 2013;**32**:853-866

[7] Manjunatha G, Lokesh V, Neelwarne B. Nitric oxide in fruit ripening: Trends and opportunities. Biotechnology Advances. 2010;**28**:489-499

[8] Lozano-Juste J, León J. Nitric oxide regulates DELLA content and PIF expression to promote photomorphogenesis in *Arabidopsis*. Plant Physiology 2011;**156**:1410-1423

[9] Terrile MC, París R, Calderón-Villalobos LI, Iglesias MJ, Lamattina L, Estelle M, Casalongué CA. Nitric oxide influences auxin signaling through S-nitrosylation of the *Arabidopsis* TRANSPORT INHIBITOR RESPONSE 1 auxin receptor. The Plant Journal. 2012;**70**:492-500

[10] Feng J, Wang C, Chen Q, Chen H, Ren B, Li X, Zuo J. S-nitrosylation of phosphotransfer proteins represses cytokinin signaling. Nature Communications. 2013;**4**:1529

[11] Freschi L. Nitric oxide and phytohormone interactions: Current status and perspectives. Frontiers in Plant Science. 2013;**4**:1-22

[12] Tossi V, Cassia R, Bruzzone S, Zocchi E, Lamattina L. ABA says NO to UV-B: A universal response?. Trends in Plant Science. 2012;**17**:510-517

[13] Zhang M, Yuan B, Leng P. The role of ABA in triggering ethylene biosynthesis and ripening of tomato fruit. Journal of Experimental Botany. 2009;**60**:1579-1588

[14] Lozano-Juste J, Leon J. Enhanced abscisic acid-mediated responses in nia1nia2noa1-2 triple mutant impaired in NIA/NR- and AtNOA1-dependent nitric oxide biosynthesis in *Arabidopsis*. Plant Physiology. 2010;**152**:891-903

[15] Lozano-Juste J, León J. Nitric oxide modulates sensitivity to ABA. Plant Signaling & Behavior. 2010;**5**:314-316.

[16] Bright J, Desikan R, Hancock JT, Weir IS, Neill SJ. ABA-induced NO generation and stomatal closure in *Arabidopsis* are dependent on H_2O_2 synthesis. The Plant Journal. 2006;**45**:113-122

[17] Lu G, Gao C, Zheng X, Han B. Identification of OsbZIP72 as a positive regulator of ABA response and drought tolerance in rice. Planta. 2009;**229**:605-615

[18] Zhang A, Jiang M, Zhang J, Ding H, Xu S, Hu X, Tan M. Nitric oxide induced by hydrogen peroxide mediates abscisic acid-induced activation of the mitogen-activated protein kinase cascade involved in antioxidant defense in maize leaves. The New Phytologist. 2007;**175**:36-50

[19] Desikan R, Griffiths R, Hancock J, Neill S. A new role for an old enzyme: Nitrate reductase-mediated nitric oxide generation is required for abscisic acid-induced stomatal closure in *Arabidopsis thaliana*. Proceedings of National Academy of Sciences. 2002;**99**:16314-16318

[20] Dubovskaya LV, Bakakina YS, Kolesneva EV, Sodel DL, McAinsh MR, Hetherington AM, Volotovski ID. cGMP-dependent ABA-induced stomatal closure in the ABA-insensitive *Arabidopsis* mutant *abi1-1*. The New Phytologist. 2011;**191**:57-69

[21] Mioto PT, Mercier H. Abscisic acid and nitric oxide signaling in two different portions of detached leaves of Guzmania monostachia with CAM up-regulated by drought. Journal of Plant Physiology. 2013;**170**:996-1002

[22] Neill S, Barros R, Bright J, Desikan R, Hancock J, Harrison J, Morris P, Ribeiro D, Wilson I. Nitric oxide, stomatal closure, and abiotic stress. Journal of Experimental Botany. 2008; **59**:165-176

[23] Sang J, Jiang M, Lin F, Xu S, Zhang A, Tan M. Nitric oxide reduces hydrogen peroxide accumulation involved in water stress-induced subcellular Anti-oxidant defense in maize plants. Journal of Integrative Plant Biology. 2008;**50**:231-243

[24] Freschi L, Rodrigues MA, Domingues DS, Purgatto E, Van Sluys MA, Magalhaes JR, Kaiser WM, Mercier H. Nitric oxide mediates the hormonal control of Crassulacean acid metabolism expression in young pineapple plants. Plant Physiology. 2010;**152**:1971-1985

[25] León J, Castillo MC, Coego A, Lozano-Juste J, Mir R. Diverse functional interactions between nitric oxide and abscisic acid in plant development and responses to stress. Journal of Experimental Botany. 2014;**65**:907-921

[26] Zhang A, Zhang J, Zhang J, Ye N, Zhang H, Tan M, Jiang M. Nitric oxide mediates brassinosteroid-induced ABA biosynthesis involved in oxidative stress tolerance in maize leaves. Plant and Cell Physiology. 2011;**52**:181-192

[27] Saito N, Yoshimasa N, Mori IC, Murata Y. Nitric oxide functions in both methyl jasmonate signaling and abscisic acid signaling in *Arabidopsis* guard cells. Plant Signaling & Behavior. 2009;**4**:119-120

[28] Liu Y, Ye N, Liu R, Chen M, Zhang J. H_2O_2 mediates the regulation of ABA catabolism and GA biosynthesis in *Arabidopsis* seed dormancy and germination. Journal of Experimental Botany. 2010;**61**:2979-2990

[29] Sanz L, Albertos P, Mateos I, Sánchez-Vicente I, Lechón T, Fernández-Marcos M, Lorenzo O. Nitric oxide (NO) and phytohormones crosstalk during early plant development. Journal of Experimental Botany. 2015;**66**:2857-2868

[30] Sugawara S, Mashiguchi K, Tanaka K, Hishiyama S, Sakai T, Hanada K, Kinoshita-Tsujimura K, Yu H, Dai X, Takebayashi Y, Takeda-Kamiya N. Distinct characteristics of indole-3-acetic acid and phenylacetic acid, two common auxins in plants. Plant and Cell Physiology. 2015;**56**:1641-1654

[31] Kohli A, Sreenivasulu N, Lakshmanan P, Kumar PP. The phytohormone crosstalk paradigm takes center stage in understanding how plants respond to abiotic stresses. Plant Cell Reports. 2013;**32**:945-957

[32] Pagnussat GC, Simontacchi M, Puntarulo S, Lamattina L. Nitric oxide is required for root organogenesis. Plant Physiology. 2002;**129**:954-956

[33] Liao WB, Huang GB, Yu JH, Zhang ML. Nitric oxide and hydrogen peroxide alleviate drought stress in marigold explants and promote its adventitious root development. Plant Physiology and Biochemistry. 2012;**58**:6-15

[34] Hunt PW, Klok EJ, Trevaskis B, Watts RA, Ellis MH, Peacock WJ, Dennis ES. Increased level of hemoglobin 1 enhances survival of hypoxic stress and promotes early growth in *Arabidopsis thaliana*. Proceeding of the National Academy of Sciences. 2002;**99**:17197-17202

[35] Chen J, Zhang HQ, Hu LB, Shi ZQ. Microcystin-LR-induced phytotoxicity in rice crown root is associated with the cross-talk between auxin and nitric oxide. Chemosphere. 2013;**93**:283-293

[36] Huang D, Wu W, Abrams SR, Cutler AJ. The relationship of drought-related gene expression in *Arabidopsis thaliana* to hormonal and environmental factors. Journal of Experimental Botany. 2008;**59**:2991-3007

[37] Munemasa S, Oda K, Watanabe-Sugimoto M, Nakamura Y, Shimoishi Y, Murata Y. The coronatine-insensitive 1 mutation reveals the hormonal signaling interaction between abscisic acid and methyl jasmonate in *Arabidopsis* guard cells. Specific impairment of ion channel activation and second messenger production. Plant Physiology. 2007;**143**:1398-1407

[38] Hossain KK, Nakamura T, Yamasaki H. Effect of nitric oxide on leaf non-photochemical quenching of fluorescence under heat-stress conditions. Russian Journal of Plant Physiology. 2011;**58**:629-633

[39] Munemasa S, Hossain MA, Nakamura Y, Mori IC, Murata Y. The *Arabidopsis* calcium-dependent protein kinase, CPK6, functions as a positive regulator of methyl jasmonate signaling in guard cells. Plant Physiology. 2011;**155**:553-561

[40] Palmieri MC, Sell S, Huang X, Scherf M, Werner T, Durner J, Lindermayr C. Nitric oxide-responsive genes and promoters in *Arabidopsis thaliana*: A bioinformatics approach. Journal of Experimental Botany. 2008;**59**:177-186

[41] Alavi SM, Arvin MJ, Manoochehri Kalantari K. Salicylic acid and nitric oxide alleviate osmotic stress in wheat (*Triticum aestivum* L.) seedlings. Journal of Plant Interactions. 2014;**9**:683-688

[42] Shao R, Wang K, Shangguan Z. Cytokinin-induced photosynthetic adaptability of Zea mays L. to drought stress associated with nitric oxide signal: Probed by ESR spectroscopy and fast OJIP fluorescence rise. Journal of Plant Physiology. 2010;**167**:472-479

[43] Xiao-Ping S, Xi-Gui S. Cytokinin and auxin-induced stomatal opening is related to the change of nitric oxide levels in guard cells in broad bean. Physiology Planta. 2006; **128**:569-579

[44] Wilhelmova N, Fuksova H, Srbova M, Mikova D, Mýtinová Z, Prochazkova D, Vytášek R, Wilhelm J. The effect of plant cytokinin hormones on the production of ethylene, nitric oxide, and protein nitrotyrosine in ageing tobacco leaves. Biofactors. 2006;**27**:203-211

[45] Liu WZ, Kong DD, Gu XX, Gao HB, Wang JZ, Xia M, Gao Q, Tian LL, Xu ZH, Bao F, Hu Y. Cytokinins can act as suppressors of nitric oxide in *Arabidopsis*. Proceedings of the National Academy of Sciences. 2013;**110**:1548-1553

[46] Rascio N, Navari-Izzo F. Heavy metal hyperaccumulating plants: How and why do they do it? And what makes them so interesting?. Plant Science. 2011;**180**:169-181

[47] Potters G, Pasternak TP, Guisez Y, Palme KJ, Jansen MA. Stress-induced morphogenic responses: Growing out of trouble?. Trends in Plant Science. 2007;**12**:98-105

[48] Piotrowska-Niczyporuk A, Bajguz A, Zambrzycka E, Godlewska-Żyłkiewicz B. Phytohormones as regulators of heavy metal biosorption and toxicity in green alga *Chlorella vulgaris* (Chlorophyceae). Plant Physiology and Biochemistry. 2012;**52**:52-65

[49] Lin YC, Chang-Chien GP, Chiang PC, Chen WH, Lin YC. Multivariate analysis of heavy metal contaminations in seawater and sediments from a heavily industrialized harbor in Southern Taiwan. Marine Pollution Bulletin. 2013;**76**:266-275

[50] He HY, He LF, Gu MH, Li XF. Nitric oxide improves aluminum tolerance by regulating hormonal equilibrium in the root apices of rye and wheat. Plant Science. 2012;**183**:123-130

[51] Thao NP, Khan MI, Thu NB, Hoang XL, Asgher M, Khan NA, Tran LS. Role of ethylene and its cross talk with other signaling molecules in plant responses to heavy metal stress. Plant Physiology. 2015;**169**:73-84

[52] Chmielowska-Bąk J, Gzyl J, Rucińska-Sobkowiak R, Arasimowicz-Jelonek M, Deckert J. The new insights into cadmium sensing. Frontiers in Plant Science. 2014;**5**:245

[53] Arasimowicz-Jelonek M, Floryszak-Wieczorek J, Gwóźdź EA. The message of nitric oxide in cadmium challenged plants. Plant Science. 2011;**181**:612-620

[54] Singh HP, Batish DR, Kaur G, Arora K, Kohli RK. Nitric oxide (as sodium nitroprusside) supplementation ameliorates Cd toxicity in hydroponically grown wheat roots. Environmental and Experimental Botany. 2008;**63**:158-167

[55] Yu Y, Li ZH, Xing DA. Nitric oxide promotes MPK6-mediated caspase-3-like activation in cadmium-induced *Arabidopsis thaliana* programmed cell death. Plant, Cell & Environment. 2013;**36**:1-5

[56] Zhao FY, Hu F, Zhang SY, Wang K, Zhang CR, Liu T. MAPKs regulate root growth by influencing auxin signaling and cell cycle-related gene expression in cadmium-stressed rice. Environmental Science and Pollution Research. 2013;**20**:5449-5460

[57] Wang Y, Loake GJ, Chu C. Cross-talk of nitric oxide and reactive oxygen species in plant programed cell death. Frontiers in Plant Science. 2013;**4**:314

[58] Chmielowska-Bąk J, Lefèvre I, Lutts S, Deckert J. Short term signaling responses in roots of young soybean seedlings exposed to cadmium stress. Journal of Plant Physiology. 2013;**170**:1585-1594

[59] Maksymiec W. Signaling responses in plants to heavy metal stress. Acta Physiologiae Plantarum. 2007;**29**:177-187

[60] Khan MI, Nazir F, Asgher M, Per TS, Khan NA. Selenium and sulfur influence ethylene formation and alleviate cadmium-induced oxidative stress by improving proline and glutathione production in wheat. Journal of Plant Physiology. 2015;**173**:9-18

[61] Thapa G, Sadhukhan A, Panda SK, Sahoo L. Molecular mechanistic model of plant heavy metal tolerance. Biometals. 2012;**25**:489-505

[62] Montero-Palmero MB, Martín-Barranco A, Escobar C, Hernández LE. Early transcriptional responses to mercury: A role for ethylene in mercury-induced stress. The New Phytologist. 2014;**201**:116-130

[63] Montero-Palmero MB, Ortega-Villasante C, Escobar C, Hernández LE. Are plant endogenous factors like ethylene modulators of the early oxidative stress induced by mercury?. Frontiers in Environmental Science 2014;**2**:34

[64] Rodríguez-serrano MA, Romero-puertas MC, Zabalza AN, Corpas FJ, Gómez M, Del Rio LA, Sandalio LM. Cadmium effect on oxidative metabolism of pea (*Pisum sativum* L.) roots. Imaging of reactive oxygen species and nitric oxide accumulation in vivo. Plant, Cell & Environment. 2006;**29**:1532-1544

[65] Cao S, Chen Z, Liu G, Jiang L, Yuan H, Ren G, Bian X, Jian H, Ma X. The *Arabidopsis* Ethylene-Insensitive 2 gene is required for lead resistance. Plant Physiology and Biochemistry. 2009;**47**:308-312

[66] Groppa MD, Rosales EP, Iannone MF, Benavides MP. Nitric oxide, polyamines and Cd-induced phytotoxicity in wheat roots. Phytochem. 2008;**69**:2609-2615

[67] Yuan HM, Huang X. Inhibition of root meristem growth by cadmium involves nitric oxide-mediated repression of auxin accumulation and signalling in *Arabidopsis*. Plant, Cell & Environment. 2016;**39**:120-135

[68] Xu J, Wang W, Yin H, Liu X, Sun H, Mi Q. Exogenous nitric oxide improves antioxidative capacity and reduces auxin degradation in roots of *Medicago truncatula* seedlings under cadmium stress. Plant and Soil 2010;**326**:321. DOI: 10.1007/s11104-009-0011-4

[69] Xu J, Wang W, Sun J, Zhang Y, Ge Q, Du L, Yin H, Liu X. Involvement of auxin and nitric oxide in plant Cd-stress responses. Plant and Soil. 2011;**346**:107

[70] Rodríguez-Serrano M, Romero-Puertas MC, Pazmino DM, Testillano PS, Risueño MC, Luis A, Sandalio LM. Cellular response of pea plants to cadmium toxicity: Cross talk between reactive oxygen species, nitric oxide, and calcium. Plant Physiology. 2009; **150**:229-243

[71] Arasimowicz-Jelonek M, Floryszak-Wieczorek J, Deckert J, Rucińska-Sobkowiak R, Gzyl J, Pawlak-Sprada S, Abramowski D, Jelonek T, Gwóźdź EA. Nitric oxide implication in cadmium-induced programmed cell death in roots and signaling response of yellow lupine plants. Plant Physiology and Biochemistry. 2012;**58**:124-134

[72] Wang Q, Liang X, Dong Y, Xu L, Zhang X, Hou J, Fan Z. Effects of exogenous nitric oxide on cadmium toxicity, element contents and antioxidative system in perennial ryegrass. Plant Growth Regulation. 2013;**69**:11-20

[73] Nahar K, Hasanuzzaman M, Alam MM, Rahman A, Suzuki T, Fujita M. Polyamine and nitric oxide crosstalk: Antagonistic effects on cadmium toxicity in mung bean plants through upregulating the metal detoxification, antioxidant defense and methylglyoxal detoxification systems. Ecotoxicology and Environmental Safety. 2016;**126**:245-255

[74] Zhu XF, Jiang T, Wang ZW, Lei GJ, Shi YZ, Li GX, Zheng SJ. Gibberellic acid alleviates cadmium toxicity by reducing nitric oxide accumulation and expression of IRT1 in *Arabidopsis thaliana*. Journal of Hazardous Materials. 2012;**239**:302-307

[75] Wu AP, Gong L, Chen X, Wang JX. Interactions between nitric oxide, gibberellic acid, and phosphorus regulate primary root growth in *Arabidopsis*. Biologia Planta. 2014;**58**:335-340

[76] Shen Q, Wang YT, Tian H, Guo FQ. Nitric oxide mediates cytokinin functions in cell proliferation and meristem maintenance in *Arabidopsis*. Molecular Plant. 2013;**6**:1214-1225

[77] Kazemi N, Khavari-Nejad RA, Fahimi H, Saadatmand S, Nejad-Sattari T. Effects of exogenous salicylic acid and nitric oxide on lipid peroxidation and antioxidant enzyme activities in leaves of *Brassica napus* L. under nickel stress. Scientia Horticulture. 2010; **126**:402-407

[78] Xu LL, Fan ZY, Dong YJ, Kong J, Bai XY. Effects of exogenous salicylic acid and nitric oxide on physiological characteristics of two peanut cultivars under cadmium stress. Biologia Planta. 2015;**59**:171-182

[79] Tossi V, Lamattina L, Cassia R. Pharmacological and genetical evidence supporting nitric oxide requirement for 2, 4-epibrassinolide regulation of root architecture in *Arabidopsis thaliana*. Plant Signaling & Behavior. 2013;**8**:e24712

[80] Choudhary SP, Kanwar M, Bhardwaj R, Yu JQ, Tran LS. Chromium stress mitigation by polyamine-brassinosteroid application involves phytohormonal and physiological strategies in *Raphanus sativus* L. PLoS One. 2012;**7**:e33210

[81] Brumbarova T, Bauer P, Ivanov R. Molecular mechanisms governing *Arabidopsis* iron uptake. Trends in Plant Science. 2015;**20**:124-133

[82] García MJ, Lucena C, Romera FJ, Alcántara E, Pérez-Vicente R. Ethylene and nitric oxide involvement in the up-regulation of key genes related to iron acquisition and homeostasis in *Arabidopsis*. Journal of Experimental Botany. 2010;**61**:3885-3899

[83] Alemayehu A, Zelinová V, Bočová B, Huttová J, Mistrík I, Tamás L. Enhanced nitric oxide generation in root transition zone during the early stage of cadmium stress is required for maintaining root growth in barley. Plant and Soil. 2015;**390**:213-222

[84] Hussain RA, Ahmad R, Nawaz F, Ashraf MY, Waraich EA. Foliar NK application mitigates drought effects in sunflower (*Helianthus annuus* L.). Acta Physiologiae Plantarum. 2016;**38**:1-4

[85] Poór P, Tari I. Ethylene-regulated reactive oxygen species and nitric oxide under salt stress in tomato cell suspension culture. Acta Biologica Szegediensis. 2011;**55**:143-146

[86] Fatma M, Masood A, Per TS, Khan NA. Nitric oxide alleviates salt stress inhibited photosynthetic performance by interacting with sulfur assimilation in mustard. Frontiers in Plant Science. 2016;**7**

[87] Molassiotis A, Tanou G, Diamantidis G. NO says more than 'YES' to salt tolerance: salt priming and systemic nitric oxide signaling in plants. Plant Signaling & Behavior. 2010;**5**:209-212

[88] Kong X, Wang T, Li W, Tang W, Zhang D, Dong H. Exogenous nitric oxide delays salt-induced leaf senescence in cotton (*Gossypium hirsutum* L.). Acta Physiologiae Plantarum. 2016;**38**:1-9

[89] Wu L, Zhang Z, Zhang H, Wang XC, Huang R. Transcriptional modulation of ethylene response factor protein JERF3 in the oxidative stress response enhances tolerance of tobacco seedlings to salt, drought, and freezing. Plant Physiology. 2008;**148**:1953-1963

[90] Wang H, Liang X, Wan Q, Wang X, Bi Y. Ethylene and nitric oxide are involved in maintaining ion homeostasis in *Arabidopsis* callus under salt stress. Planta. 2009;**230**:293-307

[91] Wang P, Du Y, Li Y, Ren D, Song CP. Hydrogen peroxide–mediated activation of MAP kinase 6 modulates nitric oxide biosynthesis and signal transduction in *Arabidopsis*. The Plant Cell. 2010;**22**:2981-2998

[92] Wang CL, Wu J, Xu GH, Gao YB, Chen G, Wu JY, Wu HQ, Zhang SL. S-RNase disrupts tip-localized reactive oxygen species and induces nuclear DNA degradation in incompatible pollen tubes of *Pyrus pyrifolia*. Journal of Cell Science. 2010;**123**:4301-4309

[93] Poór P, Gémes K, Horváth F, Szepesi A, Simon ML, Tari I. Salicylic acid treatment via the rooting medium interferes with stomatal response, CO_2 fixation rate and carbohydrate metabolism in tomato, and decreases harmful effects of subsequent salt stress. Plant Biology. 2011;**13**:105-114

[94] Liu W, Li RJ, Han TT, Cai W, Fu ZW, Lu YT. Salt stress reduces root meristem size by nitric oxide-mediated modulation of auxin accumulation and signaling in *Arabidopsis*. Plant Physiology. 2015;**168**:343-356

[95] Khokon MD, Okuma EI, Hossain MA, Munemasa S, Uraji M, Nakamura Y, Mori IC, Murata Y. Involvement of extracellular oxidative burst in salicylic acid-induced stomatal closure in *Arabidopsis*. Plant, Cell & Environment. 2011;**34**:434-443

[96] Liu N, Tu L, Tang W, Gao W, Lindsey K, Zhang X. Small RNA and degradome profiling reveals a role for miRNAs and their targets in the developing fibers of *Gossypium barbadense*. The Plant Journal. 2014;**80**:331-344

[97] Dong F, Simon J, Rienks M, Lindermayr C, Rennenberg H. Effects of rhizopheric nitric oxide (NO) on N uptake in *Fagus sylvatica* seedlings depend on soil CO_2 concentration, soil N availability and N source. Tree Physiology. 2015;**35**:910-920

[98] Khan A, Ahmad I, Shah A, Ahmad F, Ghani A, Nawaz M, Shaheen F, Fatima HU, Pervaiz F, Javed S, Hayat F. Amelioration of salinity stress in wheat (*Triticum aestivum* L) by foliar application of phosphorus. Fyton. 2013;**82**:281-287

[99] Barroso JB, Corpas FJ, Carreras A, Rodríguez-Serrano M, Esteban FJ, Fernández-Ocaña A, Chaki M, Romero-Puertas MC, Valderrama R, Sandalio LM, del Río LA. Localization of S-nitrosoglutathione and expression of S-nitrosoglutathione reductase in pea plants under cadmium stress. Journal of Experimental Botany. 2006;**57**:1785-1793

[100] Wang D, Liu Y, Tan X, Liu H, Zeng G, Hu X, Jian H, Gu Y. Effect of exogenous nitric oxide on antioxidative system and S-nitrosylation in leaves of *Boehmeria nivea* (L.) Gaud under cadmium stress. Environmental Science and Pollution Research. 2015;**22**:3489-3497

[101] Leterrier M, Chaki M, Airaki M, Valderrama R, Palma JM, Barroso JB, Corpas FJ. Function of S-nitrosoglutathione reductase (GSNOR) in plant development and under biotic/abiotic stress. Plant Signaling & Behavior. 2011;**6**:789-793

[102] Chen YS, Lo SF, Sun PK, Lu CA, Ho TH, Yu SM. A late embryogenesis abundant protein HVA1 regulated by an inducible promoter enhances root growth and abiotic stress tolerance in rice without yield penalty. Plant Biotechnology Journal. 2015;**13**:105-116

[103] Fan HF, Du CX, Guo SR. Nitric oxide enhances salt tolerance in cucumber seedlings by regulating free polyamine content. Environmental and Experimental Botany. 2013;**86**:52-59

[104] Uchida A, Jagendorf AT, Hibino T, Takabe T, Takabe T. Effects of hydrogen peroxide and nitric oxide on both salt and heat stress tolerance in rice. Plant Science. 2002;**163**:515-523

[105] Song L, Ding W, Zhao M, Sun B, Zhang L. Nitric oxide protects against oxidative stress under heat stress in the calluses from two ecotypes of reed. Plant Science. 2006;**171**:449-458

[106] Zandalinas SI, Balfagón D, Arbona V, Gómez-Cadenas A, Inupakutika MA, Mittler R. ABA is required for the accumulation of APX1 and MBF1c during a combination of water deficit and heat stress. Journal of Experimental Botany. 2016;**67**:5381-5390

[107] Daszkowska-Golec A, Szarejko I. Open or close the gate–stomata action under the control of phytohormones in drought stress conditions. Frontiers in Plant Science. 2013;**4**:138

[108] Murata Y, Mori IC, Munemasa S. Diverse stomatal signaling and the signal integration mechanism. Annual Review of Plant Biology. 2015;**66**:369-392

[109] Song L, Ding W, Shen J, Zhang Z, Bi Y, Zhang L. Nitric oxide mediates abscisic acid induced thermotolerance in the calluses from two ecotypes of reed under heat stress. Plant Science. 2008;**175**:826-832

[110] Zhou B, Guo Z, Xing J, Huang B. Nitric oxide is involved in abscisic acid-induced antioxidant activities in *Stylosanthes guianensis*. Journal of Experimental Botany. 2005;**56**:3223-3228

[111] Khan AL, Waqas M, Hamayun M, Al-Harrasi A, Al-Rawahi A, Lee IJ. Co-synergism of endophyte Penicillium resedanum LK6 with salicylic acid helped *Capsicum annuum* in biomass recovery and osmotic stress mitigation. BMC Microbiology. 2013;**13**:51

[112] Clouse SD, Sasse JM. Brassinosteroids: Essential regulators of plant growth and development. Annual Review of Plant Biology. 1998;**49**:427-451

[113] Cetinkya H, Dinler BS, Tasci E. Investigation of comparative regulation on antioxidant enzyme system under copper treatment and drought stress in maize. Notulae Botanicae Horti Agrobotanici Cluj-Napoca. 2014;**42**:363

[114] Larkindale J, Huang B. Thermotolerance and antioxidant systems in Agrostis stolonifera: Involvement of salicylic acid, abscisic acid, calcium, hydrogen peroxide, and ethylene. Journal of Plant Physiology. 2004;**161**:405-413

[115] Larkindale J, Hall JD, Knight MR, Vierling E. Heat stress phenotypes of *Arabidopsis* mutants implicate multiple signaling pathways in the acquisition of thermotolerance. Plant Physiology. 2005;**138**:882-897

[116] Kurepin LV, Qaderi MM, Back TG, Reid DM, Pharis RP. A rapid effect of applied brassinolide on abscisic acid concentrations in *Brassica napus* leaf tissue subjected to short-term heat stress. Plant Growth Regulation. 2008;**55**:165-167

[117] Bajguz A. Brassinosteroid enhanced the level of abscisic acid in *Chlorella vulgaris* subjected to short-term heat stress. Journal of Plant Physiology. 2009;**166**:882-886

[118] Atıcı Ö, Nalbantoğlu B. Antifreeze proteins in higher plants. Phytochem. 2003;**64**: 1187-1196

[119] Corpas FJ, Leterrier M, Valderrama R, Airaki M, Chaki M, Palma JM, Barroso JB. Nitric oxide imbalance provokes a nitrosative response in plants under abiotic stress. Plant Science. 2011;**181**:604-611

[120] Wimalasekera R, Villar C, Begum T, Scherer GF. COPPER AMINE OXIDASE1 (CuAO1) of *Arabidopsis thaliana* contributes to abscisic acid-and polyamine-induced nitric oxide biosynthesis and abscisic acid signal transduction. Molecular Plant. 2011;**4**:663-678

[121] Diao Q, Song Y, Shi D, Qi H. Interaction of polyamines, abscisic acid, nitric oxide, and hydrogen peroxide under chilling stress in tomato (*Lycopersicon esculentum* Mill.) seedlings. Frontiers Plant Science. 2017;**8**

[122] Esim N, Atici O. Nitric oxide improves chilling tolerance of maize by affecting apoplastic antioxidative enzymes in leaves. Plant Growth Regulation. 2014;**72**:29-38

[123] Guo Q, Wu B, Chen W, Zhang Y, Wang J, Li X. Effects of nitric oxide treatment on the cell wall softening related enzymes and several hormones of papaya fruit during storage. Revista de Agaroquimica y Tecnologia de Alimentos. 2014;**20**:309-317

[124] Li Y, Van den Ende W, Rolland F. Sucrose induction of anthocyanin biosynthesis is mediated by DELLA. Molecular Plant. 2014;**7**:570-572. DOI: 10.1093/mp/sst161

[125] Yamamoto E, Yonemaru JI, Yamamoto T, Yano M. OGRO: The overview of functionally characterized genes in rice online database. Rice. 2012;**5**:26

[126] Wang Y, Luo Z, Mao L, Ying T. Contribution of polyamines metabolism and GABA shunt to chilling tolerance induced by nitric oxide in cold-stored banana fruit. Food Chemistry. 2016;**197**:333-339

[127] Guo FQ, Okamoto M, Crawford NM. Identification of a plant nitric oxide synthase gene involved in hormonal signaling. Science. 2003;**302**:100-103

[128] Aydın B, Nalbantoğlu B. Effects of cold and salicylic acid treatments on nitrate reductase activity in spinach leaves. Turkish Journal of Biology. 2011;**35**:443-448

[129] Liu Y, Yang X, Zhu S, Wang Y. Postharvest application of MeJA and NO reduced chilling injury in cucumber (*Cucumis sativus*) through inhibition of H_2O_2 accumulation. Postharvest Biology and Technology. 2016;**119**:77-83

[130] Fu J, Chu X, Sun Y, Miao Y, Xu Y, Hu T. Nitric oxide mediates 5-aminolevulinic acid-induced antioxidant defense in leaves of *Elymus nutans* griseb. Exposed to chilling stress. PLoS One. 2015;**10**:e0130367

[131] Tamaoki M. The role of phytohormone signaling in ozone-induced cell death in plants. Plant Signaling and Behavior. 2008;**3**:166-174

[132] Ahlfors R, Brosché M, Kollist H, Kangasjärvi J. Nitric oxide modulates ozone-induced cell death, hormone biosynthesis and gene expression in *Arabidopsis thaliana*. The Plant Journal. 2009;**58**:1-2

[133] Wilkinson S, Davies WJ. Drought, ozone, ABA and ethylene: New insights from cell to plant to community. Plant, Cell & Environment. 2010;**33**:510-525

[134] Kwak JM, Nguyen V, Schroeder JI. The role of reactive oxygen species in hormonal responses. Plant Physiology. 2006;**141**:323-329

[135] Ederli L, Morettini R, Borgogni A, Wasternack C, Miersch O, Reale L, Ferranti F, Tosti N, Pasqualini S. Interaction between nitric oxide and ethylene in the induction of alternative oxidase in ozone-treated tobacco plants. Plant Physiology. 2006;**142**:595-608

[136] Rao MV, Davis KR. The physiology of ozone induced cell death. Planta. 2001;**213**:682-690

[137] Overmyer K, Tuominen H, Kettunen R, Betz C, Langebartels C, Sandermann H, Kangasjärvi J. Ozone-sensitive *Arabidopsis rcd1* mutant reveals opposite roles for ethylene and jasmonate signaling pathways in regulating superoxide-dependent cell death. The Plant Cell. 2000;**12**:1849-1862

[138] Zlatev ZS, Lidon FJ, Kaimakanova M. Plant physiological responses to UV-B radiation. Emirates Journal of Food and Agriculture. 2012;**24**:481

[139] Hideg É, Jansen MA, Strid Å. UV-B exposure, ROS, and stress: Inseparable companions or loosely linked associates?. Trends in Plant Science. 2013;**18**:107-115

[140] Wilson ID, Neill SJ, Hancock JT. Nitric oxide synthesis and signalling in plants. Plant, Cell & Environment. 2008;**31**:622-631

[141] Gonugunta VK, Srivastava N, Raghavendra AS. Cytosolic alkalinization is a common and early messenger preceding the production of ROS and NO during stomatal closure by variable signals, including abscisic acid, methyl jasmonate and chitosan. Plant Signaling & Behavior. 2009;**4**:561-564

[142] He J, Yue X, Wang R, Zhang Y. Ethylene mediates UV-B-induced stomatal closure via peroxidase-dependent hydrogen peroxide synthesis in *Vicia faba* L. Journal of Experimental Botany. 2011;**62**:2657-2666

[143] Esringu A, Aksakal O, Tabay D, Kara AA. Effects of sodium nitroprusside (SNP) pretreatment on UV-B stress tolerance in lettuce (*Lactuca sativa* L.) seedlings. Environmental Science and Pollution Research. 2016;**23**:589-597

[144] Yan F, Liu Y, Sheng H, Wang Y, Kang H, Zeng J. Salicylic acid and nitric oxide increase photosynthesis and antioxidant defense in wheat under UV-B stress. Biologia Planta. 2016;**60**:686-694

[145] Arasimowicz M, Floryszak-Wieczorek J, Milczarek G, Jelonek T. Nitric oxide, induced by wounding, mediates redox regulation in pelargonium leaves. Plant Biology. 2009;**11**:650-663

[146] Lamattina L, García-Mata C, Graziano M, Pagnussat G. Nitric oxide: The versatility of an extensive signal molecule. Annual Review of Plant Biology. 2003;**54**:109-136

[147] Wendehenne D, Durner J, Klessig DF. Nitric oxide: A new player in plant signalling and defence responses. Current Opinion in Plant Biology. 2004;**7**:449-455

[148] París R, Lamattina L, Casalongué CA. Nitric oxide promotes the wound-healing response of potato leaflets. Plant Physiology and Biochemistry. 2007;**45**:80-86

[149] Imanishi S, Kito-Nakamura K, Matsuoka K, Morikami A, Nakamura K. A major jasmonate-inducible protein of sweet potato, ipomoelin, is an ABA-independent wound-inducible protein. Plant & Cell Physiology. 1997;**38**:643-652

[150] Jih PJ, Chen YC, Jeng ST. Involvement of hydrogen peroxide and nitric oxide in expression of the ipomoelin gene from sweet potato. Plant Physiology. 2003;**132**:381-389

[151] Huang X, Stettmaier K, Michel C, Hutzler P, Mueller MJ, Durner J. Nitric oxide is induced by wounding and influences jasmonic acid signaling in *Arabidopsis thaliana*. Planta. 2004;**218**:938-946

[152] Grün S, Lindermayr C, Sell S, Durner J. Nitric oxide and gene regulation in plants. Journal of Experimental Botany. 2006;**57**:507-516

[153] Orozco-Cárdenas ML, Ryan CA. Nitric oxide negatively modulates wound signaling in tomato plants. Plant Physiology. 2002;**130**:487-493

[154] Smart CM. Gene expression during leaf senescence. The New Phytologist. 1994;**126**: 419-448

[155] Gan S, Amasino RM. Inhibition of leaf senescence by autoregulated production of cytokinin. Science. 1995;**270**:1986

[156] Grbić V, Bleecker AB. Ethylene regulates the timing of leaf senescence in *Arabidopsis*. The Plant Journal. 1995;**8**:595-602

[157] Hung KT, Kao CH. Hydrogen peroxide is necessary for abscisic acid-induced senescence of rice leaves. Journal of Plant Physiology. 2004;**161**:1347-1357

[158] Ji Y, Liu J, Xing D. Low concentrations of salicylic acid delay methyl jasmonate-induced leaf senescence by up-regulating nitric oxide synthase activity. Journal of Experimental Botony. 2016;**67**:erw280

[159] Hung KT, Kao CH. Nitric oxide counteracts the senescence of rice leaves induced by abscisic acid. Journal of Plant Physiology. 2003;**160**:871-879

[160] Mishina TE, Lamb C, Zeier J. Expression of a nitric oxide degrading enzyme induces a senescence programme in *Arabidopsis*. Plant, Cell & Environment. 2007;**30**:39-52

[161] Manjunatha G, Gupta KJ, Lokesh V, Mur LA, Neelwarne B. Nitric oxide counters ethylene effects on ripening fruits. Plant Signaling & Behavior. 2012;**7**:476-483

Chapter 7

Phytohormonal Control over the Grapevine Berry Development

Francisca Parada, Carmen Espinoza and
Patricio Arce-Johnson

Additional information is available at the end of the chapter

http://dx.doi.org/10.5772/intechopen.68453

Abstract

Grapevine (*Vitis vinifera*) is one of the most important commercial plants since its berries are used for wine production or consumed as fresh fruit or dry fruit. Many studies have focused on berry development and have pointed out the hormonal regulation on the three phases, from early development to maturity. Grapevine fruit has been classified as non-climacteric based on the low levels of ethylene present around *véraison*, although recent evidence has suggested a role for this hormone during grape berry ripening. The control of different physiological processes depends on a complex integration between environmental cues and endogenous factors, which is mediated by a phytohormone crosstalk. In this chapter, we will focus on phytohormones, their signaling pathways, and their association to berry development in *V. vinifera*; in particular, we will refer to auxins, abscisic acid, brassinosteroids, ethylene, gibberellins, and cytokinins.

Keywords: grapevine, berry, development, hormones, auxins, abscisic acid, brassinosteroids, ethylene, gibberellins, cytokinins

1. Introduction

The genus *Vitis* comprises 60–70 species and the majority of cultivated varieties pertain to *Vitis vinifera* L. The most studied process in this species is berry development, since it has a direct economic impact for wine, fresh fruit, and raisin production [1]. Berry development consists of two sigmoidal growth periods with a lag phase in between. The first stage (phase I) is characterized by a rapid cellular division and cell enlargement, establishing the final number of cells in the berries. At the end of this period, the seed embryos and berries have been defined. This stage is accompanied by the accumulation of metabolites, such as malic, tartaric, and hydroxycinnamic acids; minerals; tannins; flavonols; and volatile compounds, all essentials for normal berry development. During the lag stage (phase II), berry growth markedly diminishes, the concentration of organic acids reaches their highest level, and berries start to lose the chlorophyll accumulated during phase I. Finally, ripening stage (phase III) begins with the berry softening or *véraison*, and a gradual increase in sugar content occurs. Moreover, aroma and flavor compounds are accumulated during this stage, and anthocyanin accumulation takes place in red varieties. Also, berries increase their size by cell expansion without cellular division [2–4].

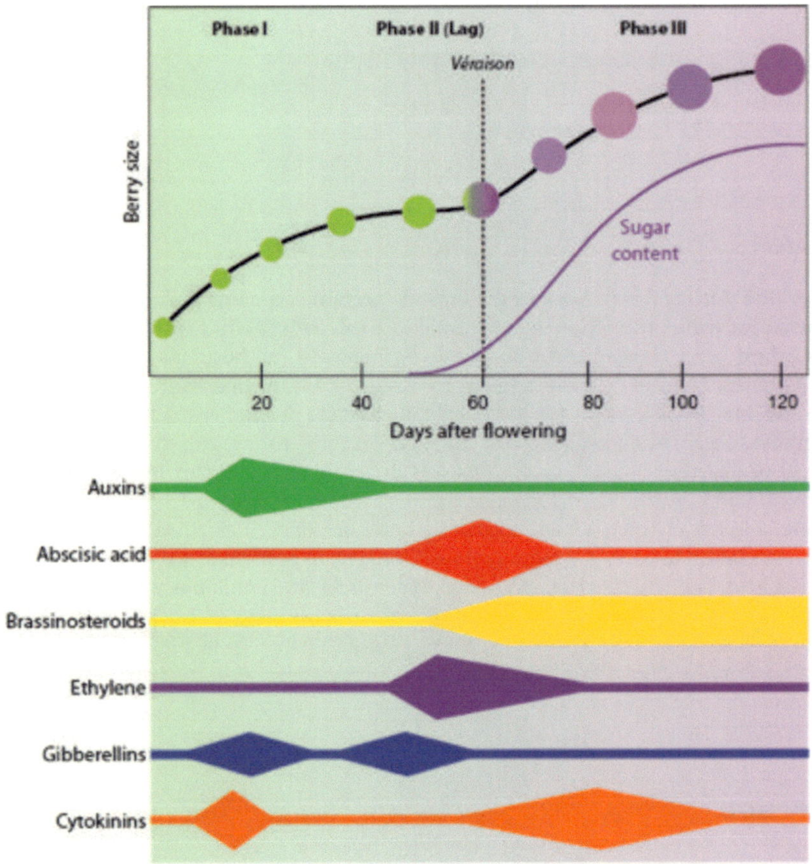

Figure 1. Schematic representation of hormonal content in grape berry development stages. Changes in hormone levels are shown for auxins, abscisic acid, brassinosteroids, ethylene, gibberellins, and cytokinins at phases I, II, and III of grape berry development. Several studies have shown a main role for auxins, cytokinins, and gibberellins in early phases of development associated to cell division process and fruit set. At *véraison*, there is a marked increase in abscisic acid levels, essential for ripening-associated physiological changes. High levels of brassinosteroids at *véraison* and phase III are consistent with a ripening-promoting role for this hormone. Moreover, a peak in ethylene has been detected prior to *véraison* that suggests a role in the initiation of this stage. Recently, a possible role for cytokinin signaling in phase III has been reported, due to its high levels at maturation. Changes in sugar content are depicted in purple curve, with an increase around *véraison* and its maximum in full mature berries. *Véraison* is denoted by a dashed line. For more details, see the text.

All the developmental phases of grapevine berry and the metabolic changes associated with them are tightly controlled by complex interactions between environmental factors, such as temperature [5–7], light [8–10], UV-B radiation [11–14], and water availability [15, 16], and endogenous factor, such as phytohormones [17–21]. Classically, fruits have been classified into climacteric and non-climacteric, where the first ones correspond to tomatoes, bananas, avocados, apples, and others that show a marked peak of ethylene associated with an increase in respiration rate during the onset of ripening. In opposition, non-climacteric fruits such as strawberry, citrus, and grapevines do not exhibit this correlation between ethylene and respiration [22–24]. However, recent evidence has suggested an unexpected role for ethylene in non-climacteric fruits [25]. In this chapter, we will discuss the current understanding of hormonal influence over berry development in grapevine, from early processes to ripening stage. In particular, the role of auxins, abscisic acid, brassinosteroids, ethylene, gibberellins, and cytokinins and its complex interaction network will be analyzed according to classic evidences and recent advances based on transcriptomic and proteomic approaches. The hormone levels at different stages of berry development are summarized in **Figure 1**.

2. Hormonal control over berry development

2.1. Auxins

Auxins are involved in several plant physiological processes, such as cell elongation and differentiation, responses to pathogen and abiotic stresses, and gravitropic and light responses, among others [26]. Indole-3-acetic acid (IAA), the main auxin in higher plants, has an essential role in initial stages of berry development. The IAA highest concentration has been observed in flowers and young berries, and it gradually decreases to a minimum low level at *véraison* [27]. Considering the high rate of cell division described during the first stage of berry development, these results are consistent with the role of auxin on this cellular process [4, 28]. The fruit set is the changeover that takes place in the ovary of a mature flower that includes the decision to abort or continue with the development of a functional fruit [29]. In grapevines, this process seems to be controlled by a hormonal balance between auxin and gibberellin (GA) [17]. Auxin induces the generation of seedless berries by parthenocarpy, a process where fruits are formed in the absence of fertilization. It has been described that genes encoding for negative regulators of fruit set initiation, AUX-IAA protein *VvIAA9* and auxin response factor (ARF) *VvARF7*, decrease their expression level during parthenocarpic berry process in cv. Tamnara treated with GA at pre-bloom stage [30]. Exogenous application of 4-chlorophenoxyacetic acid (4-CPA), a compound chemically similar to auxins, in ovaries of Fenghou grapevine, induced fruit set, but it depends on subsequent biosynthesis of gibberellin GA3 [31]. The GA signaling could be integrated to auxins by a DELLA protein mediation, since GA application increases expression of a DELLA gene [30] and the inhibition of gibberellin biosynthesis and auxin responses affected the normal physiological processes associated with fruit set [17]. In grapevine, fruit abscission occurs within 3 weeks after flowering [32], and differences in berry abscission are dependent on polar auxin transport. Experimental evidence showed a negative correlation between abscission rate and polar auxin transport, suggesting that IAA maintains the "first berries" in the cluster, i.e., those berries derived from flowers that opened the same day that flowering starts [33]. IAA delays fruitlet abscission, preventing the formation of the abscission zone by reducing its ethylene sensitivity [34]. Abscission requires a continuous polar IAA transport to the pedicel, which is controlled by IAA itself through changes in the expression and localization of PIN-FORMED (PIN) auxin efflux proteins [35]. A lower polar auxin transport rate in excised fruitlets and a decrease in the expression of *VvPINs* were correlated with an increase in abscission [35]. Taken together, these results show a key role for auxins during abscission process in grapevine.

The most studied process in berry development is the ripening. It is characterized by global transcriptomic and hormonal changes [3, 18]. Physiological changes associated with this stage have been attributed to a proper balance between ABA and auxin levels [18, 36]. IAA content is high from anthesis to *véraison* and then declines at maturation [19]. Nevertheless, there is some discrepancy about IAA variations during grapevine berry development, since no significant changes in IAA content throughout phases I and II of berry development have also been reported [37]. The maintenance of IAA low levels has been explained by a hormone inactivation mechanism, where an increase of aspartic acid-IAA conjugated form (IAA-Asp) occurs during ripening. Since this form is inactive, no biologically active IAA form will be present at this stage [38]. This is consistent with the increase in *Gretchen Hagen 1* (*GH3-1*) gene expression at *véraison*, which encodes for an IAA-amido synthetase that conjugates IAA, preferentially to aspartic acid and tryptophan [38]. Also, grape berries treated with the synthetic auxin 1-naphthaleneacetic acid (NAA) show an upregulation of GH3 gene expression [36]. In agreement to this, transcriptomic analysis using Affymetrix platform using clusters of cv. Pinot Noir collected at different development stages showed a repression in the expression of two auxin response factors (ARFs) and an auxin receptor during ripening [39]. Treatments with NAA in grape berries downregulated different metabolic pathways, such as carbohydrates, lipids, cell wall metabolism, secondary metabolites, and amino acids, and upregulated light reaction pathways [36]. This suggests that auxins might delay ripening by a negative regulation of several essential metabolic processes for normal berry maturation. Exogenous IAA application delayed the ripening process, and treatment of pre-*véraison* berries with NAA diminished the accumulation of total soluble solids (TSS) and anthocyanins levels [40–42]. Dipping of cv. Shiraz berries in a synthetic auxin-like compound benzothiazole-2-oxyacetic acid (BTOA) solution decreased the expression of chalcone synthase (*CHS*) gene of flavonoid biosynthesis pathway prior to *véraison* [43]. Moreover, BTOA caused a 2-week retardment in the onset of ripening, measured as a delay of normal increase in berry weight, color, deformability, hexose concentration, and abscisic acid content [43]. All this evidence indicates a repressor role for auxin in the typical ripening-associated physiological processes. Spraying berries with NAA delays the increase in berry size post-*véraison*, but these fruits were larger than control fruit at harvest [40]. In addition, NAA seems to delay ripening by regulation at transcriptional level, since hormone application over cv. Merlot bunches changed the expression of about 1500 genes in berries [36]. Moreover, application of NAA downregulated expression of genes encoding for 9-cis-epoxycarotenoid dioxygenase 3 (NCED3) and abscisic acid insensitive 3 (ABI3) involved in ABA biosynthesis and perception, respectively [36]. Also, NAA strongly induces the expression of genes involved in processes of conjugation, transport, and signal transduction of auxin. Particularly, NAA treatment upregulates genes coding for an Indole-3-acetic acid amido synthetase (GH3-like), protein transport inhibitor response 1 (TIR1-like), protein PIN-FORMED for polar transport (PIN3), two AUX/IAA proteins (IAA4-like and IAA31-like) known by repressed transcription of auxin-responsive genes, and an auxin response factor 8 (ARF8) [36]. On the other hand, an antagonistic relationship between auxin and ethylene has been observed. Genes encoding key enzymes of ethylene biosynthesis were upregulated in NAA-treated berries during *véraison*, as observed for 1-aminocyclopropane-1-carboxylic acid (ACC) synthase (ACS6) and ACC oxidase (ACO2) [36]. Also, genes encoding ethylene receptors, such as ethylene insensitive 4 (EIN4-like) and ethylene response sensor 1 (ERS1-like), showed an earlier increase in untreated berries compared with NAA-treated [36]. Related to the regulation of ethylene response, three genes encoding for ethylene response factors (ERF3-like, ERF-AP2-like, and ERF5-1) increase their expression in response to NAA [36]. This evidence illustrates the complex network of transcriptional responses regulating ripening process in grapevine berries.

2.2. Abscisic acid

Abscisic acid (ABA) has shown to regulate several developmental processes in plants, such as seed germination, dormancy, floral induction, and responses to environmental stresses [44]. ABA levels have been directly associated with changes in expression of the marker gene 9-cis-epoxy-carotenoid dioxygenases (NCEDs), coding for the rate-limiting enzyme in ABA biosynthesis [15, 21, 45]. Moreover, ABA can be inactivated in several ways that include conjugation with glucose or by hydroxylation into phaseic acid (PA) and dihydrophaseic acid (DPA) [44]. Thus, the active ABA content depends on the balance between biosynthesis and inactivation. Experimental evidence strongly suggests that ABA is the main signal for ripening induction in grapevine [46]. ABA levels in grape berries gradually and strongly increase just before *véraison*, accumulating in berry skin during maturity [41, 43, 45]. In agreement with this, *ABI3* expression, a transcription factor involved in ABA signaling, rises during lag phase of berry development [16, 44, 47]. Also, a correlation between the increase in anthocyanin amount and ABA content has been observed during ripening in red varieties [18, 21, 48]. An increase in ABA concentration accompanied with an upregulation of anthocyanin biosynthetic genes and anthocyanin accumulation in the skin during grape berry ripening has been reported in cv. Cabernet Sauvignon ([49] and references therein). Consistently, spraying synthetic ABA in cv. Cabernet Sauvignon immature berries at fruit set promotes anthocyanin accumulation in fruits [50]. Moreover, ABA enhances anthocyanins levels in the berry skin of cv. Cabernet Sauvignon together with an increase in *VvMYBA1* mRNA accumulation, a positive regulator of anthocyanin biosynthesis [42]. Application of ABA at *véraison* stage in cv. Crimson Seedless accelerates coloring process in berries, allowing an earlier harvesting and superior appearance compared with ethephon-treated or control berries [51]. Exogenous treatment with ABA in cv. Cabernet Sauvignon berries increases the expression of genes related to flavonoid biosynthesis, such as those encoding for stilbene synthases (STS) [52]. All this results are consistent with previous reports showing the accumulation of stilbenes, anthocyanins, and flavonols in the skin of cv. Cabernet Sauvignon berries, in response to ABA application [53]. A 2-DE proteomic approach revealed that ABA treatment increases the amount of different proteins before véraison. These proteins included vacuolar invertase GIN1, involved in hexose accumulation in the berry, and alcohol dehydrogenase 2 (ADH2); three proteins involved in flavonoid biosynthesis, chalcone isomerase, dihydroflavonol-4-reductase, and anthocyanidin reductase; and xyloglucan endotransglycosylase (XET) protein involved in cell wall modification [48]. ABA could regulate the synthesis of cell wall-modifying enzymes such as polygalacturonase [54]. The peak of activity of this enzyme coincides with the peak of ABA, and it is detected during color change period [54]. All the evidence is consistent with a role of ABA promoting berry ripening, regulating both structural and secondary metabolisms associated with berry development.

Interestingly, there is an increase in anthocyanin biosynthesis during ripening when cv. Merlot grapevines are grown under drought conditions in field experiments [15]. *NCED1* and *NCED2* genes, key enzymes from ABA biosynthesis pathway, were upregulated at the onset of *véraison* in water-deficient plants [15]. Drought stress increased ABA, proline, sugars, and anthocyanin concentration in cultivars such as Cabernet Sauvignon [45]. It has been proposed that the increase in ABA levels precedes the hexose accumulation [45]. Water deficit increased expression of genes coding for enzymes from ABA metabolism, including β-carotene hydroxylase (BHASE), involved in ABA precursors biosynthesis and NCED and (+)-abscisic acid 8′-hydroxylase (ABAHASE), involved in ABA oxidative catabolism [45]. The ABSCISIC ACID RESPONSE ELEMENT-BINDING FACTOR2 (AREB/ABFs) are a family of transcription factors involved in ABA-dependent gene activation and have been associated with ABA and abiotic stress signaling in different plant species [55]. In a recent work, the characterization of VvABF2, an AREB/ABF-like transcription factor from grape, was performed in order to understand its role during berry ripening in cv. Cabernet Sauvignon. The *VvABF2* expression increased during *véraison* and remains high during ripening until harvesting stage [55]. Interestingly, the overexpression of *VvABF2* in grape cells increased the content of the stilbenes resveratrol and piceid [55]. Berry softening is a process associated with ripening that initiates at stage II of berry development, and it depends on cell wall-modifying enzymes, including expansins, pectin methylesterase, pectate lyase, and xyloglucan endotransglycosylase/hydrolase ([56] and references therein). An early increase in ABA content during softening occurs and seems to be associated with a decrease in catabolism more than an increase in biosynthesis [56]. On the other hand, sugar unloading is a key process during berry development in which source organs, as leaves, deliver sugars to sink organs, as fruits [57]. The distribution of photoassimilates determines if the plant will go to vegetative or reproductive growth or it will accumulate starch as carbon reserve [58]. Foliar spraying of ABA in grapevines cv. Malbec demonstrated that this hormone increases glucose and fructose levels in berries and roots, probably functioning as a stress signal and enhancing sugar transport [58]. These changes in ABA-mediated sugar allocation could be an essential process favoring berry development. Just before *véraison* in Kyoho grapes, sucrose is transported into apoplast through specific carriers from apoplastic path [57]. The unloading process is possible by acid invertase enzymes. There are two types of acid invertases: the cell wall-bound invertase (CWI) that favors sucrose transport from source to sink cells and the soluble acid invertase (SAI) that has a role in storage and metabolism of sucrose in the vacuole [59]. An ABA-induced activation mechanism for these invertase regulations where ABA could regulate acid invertases at transcriptional, translational, and posttranslational levels, activating both the cell wall-bound and the soluble enzymes, has been proposed [59].

2.3. Brassinosteroids

Brassinosteroids (BRs) are a family of polyhydroxilated sterol derivatives that regulate several physiological processes in plants, such as cell elongation, biotic and abiotic stress responses, flowering, photomorphogenesis in darkness, and stomata development, among others [60, 61]. This phytohormone is the most recent being implicated in ripening of non-climacteric fruits [37, 62]. Transcriptomic analysis in cv. Merlot has suggested that BRs act as an early and key signal for ripening processes, perhaps by modulation of ethylene levels [36]. It has been shown that levels of castasterone, the bioactive BR, and its precursor 6-deoxo-castasterone increase at *véraison* and remain high during ripening in cv. Cabernet Sauvignon berries [37]. Moreover, the expression of *VvBR6OX* that encodes for brassinosteroid 6-oxidase that transform 6-deoxo-castasterone into bioactive castasterone increases until *véraison* suggesting that the high levels of castasterone and its precursor 6-deoxo-castasterone, detected at ripening, could be in part due to the increase in BR6OX enzyme [37]. Transcriptomic analysis of cv. Cabernet Sauvignon and Trincadeira has shown that the gene coding for VvBRI1, the putative brassinosteroid receptor insensitive BRI1, peaks at lag phase and then decreases, while *VvBR6OX1* peaks just prior to *véraison* and decreases thereafter [16, 63]. Moreover, putative gene coding for steroid 5 alpha reductase (*VvDET2*) was downregulated pre- and at *véraison* [63]. All the evidence suggests that the biosynthesis of brassinosteroid increases prior to *véraison* and is regulated by a feedback inhibition dependent on BR production. It has been reported that exogenous application of brassinosteroids promotes coloration of the berry skin in cv. Cabernet Sauvignon, suggesting an earlier ripening [37]. In agreement with this, application of brassinazole, an inhibitor of BR biosynthesis, delayed ripening [37]. Also, the application of exogenous brassinosteroid increases the total anthocyanin content in two cultivars, Yan73 and Cabernet Sauvignon [64, 65], and the full coloration of grapes occurred 7 days earlier in BR-treated samples [64]. BR has a major effect over genes of anthocyanin synthesis pathway, including flavonoid-3′-hydroxylase (*F3′H*), flavanone-3β-hydroxylase (*F3H*), flavonoid- 3′,5′-hydroxylase (*F3′5′H*), dihydroflavonol-4-reductase (*DFR*), anthocyanidin synthase (*ANS*), and UDP-Glc-flavonoid-3-O-glucosyl transferase (*UFGT*), and their expression peaks come ahead in about 14 days [64]. In addition to the previously described ABA role in sugar unloading in grape berries, the involvement of BR in this process has been recently reported [66]. Exogenous treatment of Cabernet Sauvignon berries with BR (24-epibrassinolide) increases the soluble sugar content; increases the activities of enzymes related to sugar unloading, neutral and acidic invertases, and sucrose synthase during ripening; and upregulates the expression of sucrose transporter genes [66]. In conclusion, BR has a role in promoting berry ripening, but more research is required to elucidate the role of BRs in berry development and its interaction network.

2.4. Ethylene

Ethylene is a gaseous hormone that can regulate several processes in plants, including carbon assimilation, flower and leaf senescence, germination, responses to abiotic and biotic stresses, organ abscission, and fruit ripening in climacteric species like apples, bananas, and tomatoes [67]. Classically, grapevine has been classified as a non-climacteric fruit due to the low levels of ethylene observed around *véraison*, suggesting that this hormone is not responsible for triggering ripening [46]. However, recent evidence showed that ethylene is present in high concentrations at anthesis and declines thereafter but clearly displayed a transient increase just prior *véraison*, suggesting now a role for this hormone in ripening initiation [25]. During early berry development, an increase in ethylene levels has been correlated with the onset of fruitlet abscission in cv. Chardonnay, which is consistent with a gain in ACC oxidase activity and high levels of ethylene precursor 1-aminocyclopropane-1-carboxylic acid (ACC) [68]. The positive effect of ethylene over berry abscission was previously reported in cv. Pinot Noir [69]. Transcriptomic analysis has shown that the gene coding for ACC synthase, from ethylene biosynthetic pathway, was repressed post-*véraison* [39]. Moreover, the ACC oxidase gene, involved in the last step of ethylene biosynthesis, showed an increased expression at lag phase with a peak around *véraison* [16, 39]. Treatments with low doses of ethylene stimulated grape berry expansion in Cabernet Sauvignon, and there is an increased expression of auxin-responsive factor 8 (ARF8) near to *véraison* in ethylene-treated berries that could suggest an interaction between ethylene and auxin signaling to control berry expansion process [70]. A possible function for ethylene signaling post-*véraison* has been suggested since some ethylene response factors (ERFs) were induced in ripening stage. However, other members of AP2/EREBP transcription factors were downregulated, indicating a complex regulation mechanism of ethylene signaling during ripening [39]. Ethylene has also been associated with anthocyanin production, since there is an increased expression of genes such as *CHS*, *F3H*, and *UFGT*, when cv. Cabernet Sauvignon berries were treated with 2-chloro-ethylphosphonic acid (2-CEPA), an ethylene-releasing compound [71]. Recently, a role of ethylene in cv. Cabernet Sauvignon fruit senescence has been postulated, since changes in expression of biosynthetic and signaling genes have been found at late ripening [49]. Taken together, this evidence suggests that ethylene has a more important role for berry development in non-climacteric species than the one though a few years ago.

2.5. Gibberellins

Gibberellins (GAs) are hormones that participate in leaf expansion, pollen maturation, seed germination, and induction of flowering, among others [72]. GAs increase early in the first phase of berry development but decrease to very low levels before *véraison* [18, 19, 37, 73], while a second peak of GAs has been reported at lag phase [74]. A role of this hormone in berry enlargement in cv. Centennial Seedless during stage I has been suggested, since GA3-treated berries showed an increased expression of xyloglucan endotransglycosylase (*XET*) genes, which participate in cell wall expansion [75]. A cross talk between GA3-ABA and GA3-ethylene during berry enlargement period has been reported [75]. An RNA-sequencing analysis of cv. Centennial seedless berries after GA3 application showed a decrease in negative regulators of ABA content and signaling, suggesting that GA3 promotes ABA signaling, while there is an upregulation of ethylene response factors (ERFs) at 1 and 3 days post GA3 treatment with a downregulation observed 7 days post treatment [75]. Moreover, GAs have been involved in mesocarp cell expansion, and treatment with paclobutrazol (PAC), an inhibitor of GA biosynthesis, reduces fruit set [31]. An analysis of different GA3-oxidases revealed that *VvGA3ox1*, involved in GA biosynthesis, increases its expression at fruit set [76]. It has been reported that GA3 increases glucose and fructose content in berries of cv. Malbec at *véraison*, therefore affecting sugar distribution [58]. Also, experimental results proposed that fruitlet abscission could be coregulated by levels of auxin and GA, since both reduce the polar transport of auxin [35]. On the other hand, there is not much information implicating GAs in ripening process. However, GA3 soaking of cv. Kyoho clusters strongly accelerates berry coloration compared with non-treated ones [77]. Thus, it is possible that GAs could participate in anthocyanin biosynthesis regulation, probably in a complex interaction with other hormones. A transcriptomic analysis of cv. Cabernet Sauvignon berries showed that GA β-hydroxylase gene, involved in GA biosynthesis, has its highest expression at phase I and then decreases in phases II and III [16]. Also, there is a gradual increase in gibberellin receptor *GIDL1* expression, which is more marked for *GIDL2,* as phases II and III progress [16]. These results are consistent with an increase in *GILD2* expression at *véraison* in cv. Trincadeira berries [63]. Nevertheless, more studies are needed to determine the precise function of this hormone during berry development, especially during ripening stage.

2.6. Cytokinins

Cytokinins are involved in seed germination, cell proliferation and differentiation, light responses, delayed of senescence, and others [78]. During early grapevine development, cytokinin 6-benzyladenine (6-BA) treatment induces seedless berry formation through parthenocarpy [31]. The levels of zeatin and zeatin riboside, the more active forms of cytokinins, were high in early phase I and then decrease in berry flesh to undetectable levels post-*véraison* [27]. This suggests that cytokinins do not participate in ripening stage. This is consistent with reports that exogenous application of cytokinins in cv. Delaware did not affect fructose content in berries, a key characteristic of ripening [79]. Furthermore, the expression of genes encoding for dehydrogenases, putatively involved in cytokinin degradation, decreased around *véraison* in cv. Trincadeira berries [63]. However, recent evidence indicates that cytokinins could have a role in ripening. For instance, isopentenyladenine (iP), a biological active cytokinin, increases in cv. Shiraz berries during ripening, while, during late ripening, there is a higher amount of t-ZOG, a glycosylated form of inactive cytokinin [80]. Recently, an analysis of cytokinin levels was performed in grape cultivars Cabernet Sauvignon, Riesling, and Pinot Noir, as well as in strawberry and tomato. The results indicated a markedly increase in iP concentration in red firm tomato, red ripe strawberry, and post-*véraison* grapes, based on which the hypothesis of a common role for cytokinin signaling during ripening in climacteric and non-climacteric fruits was postulated [81]. Throughout berry development, genes related to biosynthesis, catabolism, perception, and signaling of cytokinins are expressed [81]. Also, the repression of the gene encoding for cytokinin-repressed 9 protein (CR9) has been reported in ripening, supporting the activation of cytokinin signaling during this stage [39, 63]. The gene expression of nuclear regulators of cytokinin signaling (ARR) had different patterns around *véraison* in cv. Trincadeira grapes, complicating the interpretation [63]. So, it is still needed to clarify cytokinins' exact participation in phase III of berry development.

3. Conclusions

The role of the different hormones during grapevine ripening seems to be complex. Currently, some technical difficulties limit a better understanding of physiological processes occurring during grapevine berry development and the hormonal control. There is no availability of a mutant collection of *V. vinifera* that could support the study of hormonal signaling or biosynthetic pathways to clearly understand hormone control of berry formation and maturity allowing to establish direct cause-effect relationships. Transgenesis in grapevine is not an easygoing procedure, and the generation of overexpressing or silencing lines in genes of interest is a difficult task. At the moment, correlations between hormone levels and gene expression at different phases of berry development are the best approach to unravel hormonal regulation of berry development. So far, a successful approach has been exogenous treatments with hormones, which have allowed to suggest conclusions and discern the complex interaction network. In addition, global analyses including genomics, transcriptomics, and proteomics have improved our knowledge about this process. It is expected that the progress of new technologies could help to overcome technical limitations and give more insight into control of berry development mediated by plant hormones.

Acknowledgements

This work was partially supported by National Commission for Science and Technology CONICYT (FONDECYT grant number 115020), Project PMG Vides CORFO 13CTI-18862, and Millennium Nucleus for Plant Synthetic Biology and Systems Biology NC130030. F. Parada was supported by VRI doctoral grant from Pontificia Universidad Católica de Chile.

Author details

Francisca Parada, Carmen Espinoza and Patricio Arce-Johnson*

*Address all correspondence to: parce@bio.puc.cl

Department of Molecular Genetics and Microbiology, Faculty of Biological Sciences, Pontifical Catholic University of Chile, Santiago, Chile

References

[1] Keller M. The Science of Grapevines: Anatomy and Physiology. The Science of Grapevines. Academic Press: San Diego, USA; 2015. pp. 4-6. DOI: 10.1016/B978-0-12-374881-2.00012-X

[2] Dokoozlian N. Grape berry growth and development. In: Christensen P, editor. Raisin Production Manual. Oakland, CA: Agricultural and Natural Resources Publication; 2000. pp.30-37

[3] Fortes AM, Teixeira RT, Agudelo-Romero P. Complex interplay of hormonal signals during grape berry ripening. Molecules. 2015;**20**:9326-9343. DOI: 10.3390/molecules20059326

[4] Kennedy BYJ. Understanding grape berry development. Practical Winery and Vineyard. 2002;**4**:1-5. DOI: 10.1111/j.1755-0238.2000.tb00171.x

[5] Carbonell-Bejerano P, Santa María E, Torres-Pérez R, Royo C, Lijavetzky D, Bravo G, Aguirreolea J, Sánchez-Díaz M, Antolín MC, Martínez-Zapater JM. Thermotolerance responses in ripening berries of *Vitis vinifera* L. cv Muscat Hamburg. Plant and Cell Physiology. 2013;**54**:1200-1216. DOI: 10.1093/pcp/pct071

[6] Cohen SD, Tarara JM, Gambetta GA, Matthews MA, Kennedy JA. Impact of diurnal temperature variation on grape berry development, proanthocyanidin accumulation, and the expression of flavonoid pathway genes. Journal of Experimental Botany. 2012;**63**:2655-2665. DOI: 10.1093/jxb/err449

[7] Yamane T, Seok TJ, Goto-Yamamoto N, Koshita Y, Kobayashi S. Effects of temperature on anthocyanin biosynthesis in grape berry skins. American Journal of Enology and Viticulture. 2006;**57**:54-59

[8] Domingos S, Fino J, Cardoso V, Sánchez C, Ramalho JC, Larcher R, Paulo OS, Oliveira CM, Goulao LF. Shared and divergent pathways for flower abscission are triggered by gibberellic acid and carbon starvation in seedless *Vitis vinifera* L. BMC Plant Biology. 2016;**16**:38. DOI: 10.1186/s12870-016-0722-7

[9] Downey MO, Harvey JS, Robinson SP. The effect of bunch shading on berry development and flavonoid accumulation in Shiraz grapes. Australian Journal of Grape and Wine Research. 2004;**10**:55-73. DOI: 10.1111/j.1755-0238.2004.tb00008.x

[10] Matus JT, Loyola R, Vega A, Peña-Neira A, Bordeu E, Arce-Johnson P, Alcalde JA. Post-veraison sunlight exposure induces MYB-mediated transcriptional regulation of anthocyanin and flavonol synthesis in berry skins of *Vitis vinifera*. Journal of Experimental Botany. 2009;**60**:853-867. DOI: 10.1093/jxb/ern336

[11] Carbonell-Bejerano P, Diago MP, Martínez-Abaigar J, Martínez-Zapater JM, Tardáguila J, Núñez-Olivera E. Solar ultraviolet radiation is necessary to enhance grapevine fruit ripening transcriptional and phenolic responses. BMC Plant Biology. 2014;**14**:183. DOI: 10.1186/1471-2229-14-183

[12] Martinez-Luscher J. Effects of UV-B radiation on grapevine (*Vitis vinifera* cv. Tempranillo) leaf physiology and berry composition, framed within the climate change scenario (water deficit, elevated CO_2 and elevated temperature) [thesis]. Bordeaux: Université de Bordeaux; 2014

[13] Martínez-Lüscher J, Morales F, Delrot S, Sánchez-Díaz M, Gomès E, Aguirreolea J, Pascual I. Characterization of the adaptive response of grapevine (cv. Tempranillo) to UV-B radiation under water deficit conditions. Plant Science. 2015;**232**:13-22. DOI: 10.1016/j.plantsci.2014.12.013

[14] Matus JT. Transcriptomic and metabolomic networks in the grape berry illustrate that it takes more than flavonoids to fight against ultraviolet radiation. Frontiers in Plant Science. 2016;**7**:1337. DOI: 10.3389/fpls.2016.01337

[15] Castellarin SD, Pfeiffer A, Sivilotti P, Degan M, Peterlunger E, Gaspero GDI. Transcriptional regulation of anthocyanin biosynthesis in ripening fruits of grapevine

under seasonal water deficit. Plant, Cell & Environment. 2007;**30**:1381-1399. DOI: 10.1111/j.1365-3040.2007.01716.x

[16] Deluc LG, Grimplet J, Wheatley MD, Tillett RL, Quilici DR, Osborne C, Schooley DA, Schlauch KA, Cushman JC, Cramer GR. Transcriptomic and metabolite analyses of Cabernet Sauvignon grape berry development. BMC Genomics. 2007;**8**:429. DOI: 10.1186/1471-2164-8-429

[17] Godoy MF. Hormonal regulation during initial berry development in grapevine [thesis]. Bordeaux: Université de Bordeaux; 2013

[18] Böttcher C, Davies C. Hormonal control of grape berry ripening. In: Roubelakis-Angelakis KA, editor. Grapevine Molecular Physiology & Biotechnology. 2nd ed. Heraklion: Springer; 2009. pp. 229-256. DOI: 10.1007/978-90-481-2305-6

[19] Conde C, Silva P, Fontes N, Dias ACP, Tavares RM, Sousa MJ, Agasse A, Delrot S, Gerós H. Biochemical changes throughout grape berry development and fruit and wine quality. Food. 2007;**1**:1-22

[20] Kuhn N, Guan L, Dai ZW, Wu BH, Lauvergeat V, Gomès E, Li SH, Godoy F, Arce-Johnson P, Delrot S. Berry ripening: Recently heard through the grapevine. Journal of Experimental Botany. 2014;**65**:4543-4559. DOI: 10.1093/jxb/ert395

[21] Wheeler S, Loveys B, Ford C, Davies C. The relationship between the expression of abscisic acid biosynthesis genes, accumulation of abscisic acid and the promotion of *Vitis vinifera* L. berry ripening by abscisic acid. Australian Journal of Grape and Wine Research. 2009;**15**:195-204. DOI: 10.1111/j.1755-0238.2008.00045.x

[22] Burg SP, Burg EA. Role of ethylene in fruit ripening. Plant Physiology. 1962;**37**:179-189. DOI: 10.1104/pp.37.2.179

[23] Hale CR, Coombe BG, Hawker JS. Effects of ethylene and 2-chloroethylphosphonic acid on the ripening of grapes. Plant Physiology. 1970;**45**:620-623. DOI: 10.1104/pp.45.5.620

[24] Symons G, Chua YJ, Ross J, Quittenden LJ, Davies NW, Reid JB. Hormonal changes during non-climacteric ripening in strawberry. Journal of Experimental Botany. 2012;**63**:4741-4750. DOI: 10.1093/jxb/ers147

[25] Chervin C, El-Kereamy A, Roustan JP, Latché A, Lamon J, Bouzayen M. Ethylene seems required for the berry development and ripening in grape, a non-climacteric fruit. Plant Science. 2004;**167**:1301-1305. DOI: 10.1016/j.plantsci.2004.06.026

[26] Sauer M, Robert S, Kleine-Vehn J. Auxin: Simply complicated. Journal of Experimental Botany. 2013;**64**:2565-2577. DOI: 10.1093/jxb/ert139

[27] Zhang X, Luo G, Wang R, Wang J, Himelrick DG. Growth and developmental responses of seeded and seedless grape berries to shoot girdling. Journal of the American Society for Horticultural Science. 2003;**128**:316-323

[28] Perrot-Rechenmann C. Cellular responses to auxin: Division versus expansion. Cold Spring Harbor Perspectives in Biology. 2010;**2**:1-15. DOI: 10.1101/cshperspect.a001446

[29] Gillaspy G, Ben-David H, Gruissem W. Fruits: A Developmental Perspective. The Plant Cell. 1993;**5**:1439-1451. DOI: 10.1105/tpc.5.10.1439

[30] Jung CJ, Hur YY, Yu HJ, Noh JH, Park KS, Lee HJ. Gibberellin application at prebloom in grapevines down-regulates the expressions of *VvIAA9* and *VvARF7*, negative regulators of fruit set initiation, during parthenocarpic fruit development. PLoS One. 2014;**9**:e95634. DOI:10.1371/journal.pone.0095634

[31] Lu L, Liang J, Zhu X, Xiao K, Li T, Hu J. Auxin- and cytokinin-induced berries set in grapevine partly rely on enhanced gibberellin biosynthesis. Tree Genetics and Genomes. 2016;**12**:41. DOI: 10.1007/s11295-016-0980-4

[32] Intrigliolo D S, Lakso AN. Berry abscission is related to berry growth in "*Vitis labruscana*" Concord and "*Vitis vinifera*" Riesling. Vitis. 2009;**48**:53-54

[33] Kühn N, Abello C, Godoy F, Delrot S, Arce-Johnson P. Differential behavior within a grapevine cluster: Decreased ethylene-related gene expression dependent on auxin transport is correlated with low abscission of first developed berries. PloS One. 2014;**9**:e111258. DOI: 10.1371/journal.pone.0111258

[34] Bangerth F. Abscission and thinning of young fruit and their regulation by plant hormones and bioregulators. Plant Growth Regulation. 2000;**31**:43-59. DOI: 10.1023/a:1006398513703

[35] Kühn N, Serrano A, Abello C, Arce A, Espinoza C, Gouthu S, Deluc L, Arce-Johnson P. Regulation of polar auxin transport in grapevine fruitlets (*Vitis vinifera* L.) and the proposed role of auxin homeostasis during fruit abscission. BMC Plant Biology. 2016;**16**:1-17. DOI: 10.1186/s12870-016-0914-1

[36] Ziliotto F, Corso M, Rizzini FM, Rasori A, Botton A, Bonghi C. Grape berry ripening delay induced by a pre-véraison NAA treatment is paralleled by a shift in the expression pattern of auxin- and ethylene-related genes. BMC Plant Biology. 2012;**12**:185. DOI: 10.1186/1471-2229-12-185

[37] Symons GM, Davies C, Shavrukov Y, Dry IB, Reid JB, Thomas MR. Grapes on steroids. Brassinosteroids are involved in grape berry ripening. Plant Physiology. 2006;**140**:150-158. DOI: 10.1104/pp.105.070706

[38] Böttcher C, Keyzers RA, Boss PK, Davies C. Sequestration of auxin by the indole-3-acetic acid-amido synthetase GH3-1 in grape berry (*Vitis vinifera* L.) and the proposed role of auxin conjugation during ripening. Journal of Experimental Botany. 2010;**61**:3615-3625. DOI: 10.1093/jxb/erq174

[39] Pilati S, Perazzolli M, Malossini A, Cestaro A, Demattè L, Fontana P, Dal Ri A, Viola R, Velasco R, Moser C. Genome-wide transcriptional analysis of grapevine berry ripening reveals a set of genes similarly modulated during three seasons and the occurrence of an oxidative burst at vèraison. BMC Genomics. 2007;**8**:428. DOI: 10.1186/1471-2164-8-428

[40] Böttcher C, Harvey K, Forde CG, Boss PK, Davies C. Auxin treatment of pre-veraison grape (*Vitis vinifera* L.) berries both delays ripening and increases the synchronicity of

sugar accumulation. Australian Journal of Grape and Wine Research. 2011;**17**:1-8. DOI: 10.1111/j.1755-0238.2010.00110.x

[41] Deytieux-Belleau C, Gagne S, L'Hyvernay A, Doneche B, & Geny, L. Possible roles of both abscisic acid and indol-acetic acid in controlling grape berry ripening process. Journal International Des Sciences de La Vigne et Du Vin. 2007;**41**:141-148

[42] Jeong ST, Goto-Yamamoto N, Kobayashi S, Esaka M. Effects of plant hormones and shading on the accumulation of anthocyanins and the expression of anthocyanin biosynthetic genes in grape berry skins. Plant Science. 2004;**167**:247-252. DOI: 10.1016/j.plantsci.2004.03.021

[43] Davies C, Boss PK, Robinson SP. Treatment of grape berries, a nonclimacteric fruit with a synthetic auxin, retards ripening and alters the expression of developmentally regulated genes. Plant Physiology. 1997;**115**:1155-1161. DOI: 10.1104/pp.115.3.1155

[44] Finkelstein R. Abscisic acid synthesis and response. The Arabidopsis Book. 2013;**11**:e0166. DOI: 10.1199/tab.0166

[45] Deluc LG, Quilici DR, Decendit A, Grimplet J, Wheatley MD, Schlauch KA, Mérillon JM, Cushman JC, Cramer GR. Water deficit alters differentially metabolic pathways affecting important flavor and quality traits in grape berries of Cabernet Sauvignon and Chardonnay. BMC Genomics. 2009;**10**:1-33. DOI: 10.1186/1471-2164-10-212

[46] Coombe BG, Hale CR. The hormone content of ripening grape berries and the effects of growth substance treatments. Plant Physiology. 1973;**51**:629-634. DOI: 10.1104/pp.51.4.629

[47] Giraudat J, Hauge BM, Valon C, Smalle J, Parcy F, Goodman HM. Isolation of the Arabidopsis ABI3 gene by positional cloning. The Plant Cell. 1992;**4**:1251-1261. DOI: 10.1105/tpc.4.10.1251

[48] Giribaldi M, Gény L, Delrot S, Schubert A. Proteomic analysis of the effects of ABA treatments on ripening *Vitis vinifera* berries. Journal of Experimental Botany. 2010;**61**:2447-2458. DOI: 10.1093/jxb/erq079

[49] Cramer GR, Ghan R, Schlauch KA, Tillett RL, Heymann H, Ferrarini A, Delledonne M, Zenoni S, Fasoli M, Pezzotti M. Transcriptomic analysis of the late stages of grapevine (*Vitis vinifera* cv. Cabernet Sauvignon) berry ripening reveals significant induction of ethylene signaling and flavor pathways in the skin. BMC Plant Biology. 2014;**14**:370. DOI: 10.1186/s12870-014-0370-8

[50] Gagné S, Esteve K, Deytieux C, Saucier C, Geny L. Influence of abscisic acid in triggering "véraison" in grape berry skins of *Vitis vinifera* L. cv. Cabernet Sauvignon. Journal International Des Sciences de La Vigne et Du Vin. 2006;**40**:7-14

[51] Cantín CM, Fidelibus MW, Crisosto CH. Application of abscisic acid (ABA) at veraison advanced red color development and maintained postharvest quality of "Crimson Seedless" grapes. Postharvest Biology and Technology. 2007;**46**:237-241. DOI: 10.1016/j.postharvbio.2007.05.017

[52] Rattanakon S, Ghan R, Gambetta GA, Deluc LG, Schlauch KA, Cramer GR. Abscisic acid transcriptomic signaling varies with grapevine organ. BMC Plant Biology. 2016;**16**:72. DOI: 10.1186/s12870-016-0763-y

[53] Koyama K, Sadamatsu K, Goto-Yamamoto N. Abscisic acid stimulated ripening and gene expression in berry skins of the Cabernet Sauvignon grape. Functional and Integrative Genomics. 2010;**10**:367-381. DOI: 10.1007/s10142-009-0145-8

[54] Deytieux, C, Gény, L, & Doneche, B. Relation between hormonal balance and polygalacturonase activity in grape berry. Acta Horticulturae. 2005;**682**:163-170. DOI: 10.17660/ActaHortic.2005.682.15

[55] Nicolas P, Lecourieux D, Kappel C, Cluzet S, Cramer G, Delrot S, Lecourieux F. The basic leucine zipper transcription factor ABSCISIC ACID RESPONSE ELEMENT-BINDING FACTOR2 is an important transcriptional regulator of abscisic acid-dependent grape berry ripening processes. Plant Physiology. 2014;**164**:365-383. DOI: 10.1104/pp.113.231977

[56] Castellarin SD, Gambetta GA, Wada H, Krasnow MN, Cramer GR, Peterlunger E, Shackel KA, Matthews MA. Characterization of major ripening events during softening in grape: Turgor, sugar accumulation, abscisic acid metabolism, colour development, and their relationship with growth. Journal of Experimental Botany. 2016;**67**:709-722. DOI: 10.1093/jxb/erv483

[57] Zhang XY, Wang XL, Wang, XF, Xia GH, Pan QH, Fan RC, Wu FQ, Yu XC, Zhang DP. A shift of phloem unloading from symplasmic to apoplasmic pathway is involved in developmental onset of ripening in grape berry. Plant Physiology. 2006;**142**:220-232. DOI: 10.1104/pp.106.081430

[58] Moreno D, Berli FJ, Piccoli PN, Bottini R. Gibberellins and abscisic acid promote carbon allocation in roots and berries of grapevines. Journal of Plant Growth Regulation. 2011;**30**:220-228. DOI: 10.1007/s00344-010-9186-4

[59] Pan QH, Li MJ, Peng CC, Zhang N, Zou X, Zou KQ, Wang XL, Yu XC, Wang XF, Zhang DP. Abscisic acid activates acid invertases in developing grape berry. Physiologia Plantarum. 2005;**125**:157-170. DOI: 10.1111/j.1399-3054.2005.00552.x

[60] Clouse SD. Brassinosteroids. The Arabidopsis Book. 2011;**9**:e0151. DOI: 10.1199/tab.0151

[61] Zhu J, Sae-Seaw J, Wang, Z. Brassinosteroid signalling. Development. 2013;**140**:1615-1620. DOI: 10.1242/dev.060590

[62] Chai Y, Zhang Q, Tian L, Li CL, Xing Y, Qin L, Shen YY. Brassinosteroid is involved in strawberry fruit ripening. Plant Growth Regulation. 2013;**69**:63-69. DOI: 10.1007/s10725-012-9747-6

[63] Fortes AM, Agudelo-Romero P, Silva MS, Ali K, Sousa L, Maltese F, Choi YH, Grimplet J, Martinez-Zapater JM, Verpoorte R, Pais MS. Transcript and metabolite analysis in *Trincadeira cultivar* reveals novel information regarding the dynamics of grape ripening. BMC Plant Biology. 2011;**11**:149. DOI: 10.1186/1471-2229-11-149

[64] Luan L, Zhang Z, Xi Z, Huo S, Ma L. Brassinosteroids regulate anthocyanin biosynthesis in the ripening of grape berries. South African Journal of Enology and Viticulture. 2013;**34**:196-203.

[65] Xi ZM, Zhang ZW, Huo SS, Luan LY, Gao X, Ma LN, Fang YL. Regulating the secondary metabolism in grape berry using exogenous 24-epibrassinolide for enhanced phenolics content and antioxidant capacity. Food Chemistry. 2013;**141**:3056-3065. DOI: 10.1016/j.foodchem.2013.05.137

[66] Xu F, Xi ZM, Zhang H, Zhang CJ, Zhang ZW. Brassinosteroids are involved in controlling sugar unloading in *Vitis vinifera* "Cabernet Sauvignon" berries during véraison. Plant Physiology and Biochemistry. 2015;**94**:197-208. DOI: 10.1016/j.plaphy.2015.06.005

[67] Bleecker AB, Kende H. Ethylene: A gaseous signal molecule in plants. Annual Review of Cell and Developmental Biology. 2000;**16**:1-18. DOI: 10.1146/annurev.cellbio.16.1.1

[68] Hilt C, Bessis R. Abscission of grapevine fruitlets in relation to ethylene biosynthesis. Vitis. 2003;**42**:1-3.

[69] Bessis R, Charpentier N, Hilt C, Fournioux JC. Grapevine fruit set: Physiology of the abscission zone. Australian Journal of Grape and Wine Research. 2000;**6**:125-130. DOI: 10.1111/j.1755-0238.2000.tb00170.x

[70] Chervin C, Tira-Umphon A, Terrier N, Zouine M, Severac D, Roustan JP. Stimulation of the grape berry expansion by ethylene and effects on related gene transcripts, over the ripening phase. Physiologia Plantarum. 2008;**134**:534-546. DOI: 10.1111/j.1399-3054.2008.01158.x

[71] El-Kereamy A, Chervin C, Roustan JP, Cheynier V, Souquet JM, Moutounet M, Raynal J, Ford C, Latché A, Pech JC, Bouzayen M. Exogenous ethylene stimulates the long-term expression of genes related to anthocyanin biosynthesis in grape berries. Physiologia Plantarum. 2003;**119**:175-182. DOI: 10.1034/j.1399-3054.2003.00165.x

[72] Achard P, Genschik P. Releasing the brakes of plant growth: How GAs shutdown DELLA proteins. Journal of Experimental Botany. 2009;**60**:1085-1092. DOI: 10.1093/jxb/ern301

[73] Inaba A, Ishida M, Sobajima Y. Changes in endogenous hormone concentrations during berry development in relation to the ripening of delaware grapes. Journal of the Japanese Society for Horticultural Sciences. 1976;**45**:245-252. DOI: 10.2503/jjshs.45.245

[74] Pérez FJ, Viani C, Retamales J. Bioactive gibberellins in seeded and seedless grapes: Identification and changes in content during berry development. American Journal of Enology and Viticulture. 2000;**51**:315-318

[75] Chai L, Li Y, Chen S, Perl A, Zhao F, Ma H. RNA sequencing reveals high resolution expression change of major plant hormone pathway genes after young seedless grape berries treated with gibberellin. Plant Science. 2014;**229**:215-224. DOI: 10.1016/j.plantsci.2014.09.010

[76] Giacomelli L, Rota-Stabelli O, Masuero D, Acheampong AK, Moretto M, Caputi L, Vrhovsek U, Moser C. Gibberellin metabolism in *Vitis vinifera* L. during bloom and fruit-

set: Functional characterization and evolution of grapevine gibberellin oxidases. Journal of Experimental Botany. 2013;**64**:4403-4419. DOI: 10.1093/jxb/ert251

[77] Cheng C, Jiao C, Singer SD, Gao M, Xu X, Zhou Y, Li Z, Fei Z, Wang Y, Wang X. Gibberellin-induced changes in the transcriptome of grapevine (*Vitis labrusca* × *V. vinifera*) cv. Kyoho flowers. BMC Genomics. 2015;**16**:128. DOI: 10.1186/s12864-015-1324-8

[78] Hwang I, Sheen J, Uller B. Cytokinin signaling networks. Annual Review of Plant Biology. 2012;**63**:353-380. DOI: 10.1146/annurev-arplant-042811-105503

[79] Inaba A, Ishida M, Sobajima Y. Regulation of Ripening in Grapes by Hormone Treatments. Scientic Reports of Kyoto Prefectural University: Kyoto, Japan. 1974; pp. 25-31.

[80] Böttcher C, Boss PK, Davies C. Increase in cytokinin levels during ripening in developing *Vitis vinifera* cv. Shiraz berries. American Journal of Enology and Viticulture. 2013;**64**:527-531. DOI: 10.5344/ajev.2013.13053

[81] Böttcher C, Burbidge CA, Boss PK, Davies C. Changes in transcription of cytokinin metabolism and signalling genes in grape (*Vitis vinifera* L.) berries are associated with the ripening-related increase in isopentenyladenine. BMC Plant Biology. 2015;**15**:223. DOI: 10.1186/s12870-015-0611-5